ANTARCTIC AND SUBANTARCTIC SCLERACTINIA

Stephen D. Cairns

Department of Invertebrate Zoology, Smithsonian Institution
Washington, D. C. 20560

Abstract. The 37 species of Scleractinia known from the Antarctic and Subantarctic regions are described, mapped, and illustrated, including the description of 6 new species. Two more species, one of them new, from waters closely adjacent to the Subantarctic regions are also considered, as well as 4 previously published records of Scleractinia of uncertain identity. A chronological, annotated list of all papers dealing with Antarctic Scleractinia is provided. A zoogeographic analysis describes common patterns of species distributions, and the faunas of various geographic areas are discussed. Scleractinia from South Pacific seamounts, some of which may form deepwater coral banks, are particularly noted.

Introduction

This paper reviews the 37 species of Scleractinia known from the Antarctic and Subantarctic regions, as well as 4 records of uncertain identity from these regions, and 2 more species from areas closely adjacent to the Subantarctic region. Despite 28 papers dealing exclusively or partially with Antarctic Scleractinia (Table 1), only about 1200 specimens have been reported previously from 114 localities. The material upon which this study is based (primarily United States Antarctic Research Program vessels, Deep Freeze vessels, and the R/V Vema) includes over 7 times the number of specimens and almost 3 times the number of localities reported in all the previous papers.

Squires [1969] published a preliminary synthesis of the distribution of Antarctic Scleractinia but never did write his intended Antarctic monograph, thereby leaving many questions unanswered. Although his paper is valuable as a preliminary note, it includes numerous undocumented range extensions (all of his mapped records are undocumented), unexplained statements about complex groups of species, omissions of previous records, and reference to an undescribed new species. Most, if not all, of Squires's [1969] specimens are deposited at the United States National Museum, Washington, D. C., and at the American Museum of Natural History, New York. This material, along with subsequently collected United States Antarctic Research Program specimens and an examination of most of the previously reported specimens, has allowed a documentation, and sometimes a correction, of Squires's synthesis, a reevaluation of the systematics of the fauna, and a more meaningful zoogeographic analysis of the fauna.

Material and Methods

This study is based on the examination of approximately 8700 specimens collected throughout the Antarctic, Subantarctic, and adjacent waters and includes new material from 482 lots collected at 310 stations. Most of the specimens were collected by the vessels participating in the United States Antarctic Research Program (USNS Eltanin, now the ARA Islas Orcadas, and R/V Hero) and those participating in Operation Deep Freeze III and IV. These specimens, as well as those from other vessels (e.g., USCGC Eastwind, USFCS Albatross, and R/V Anton Bruun), are deposited at the United States National Museum. Other collections examined include specimens collected by the R/V Vema (deposited at the American Museum of Natural History), specimens collected by the Walther Herwig (deposited at the Zoologisches Institut und Zoologisches Museum, Hamburg), and a small collection of USNS Eltanin corals deposited at the Museum of Comparative Zoology, Harvard, Cambridge.

Previously reported specimens from the following museums were examined: British Museum, London; Muséum National d'Histoire Naturelle, Paris; Museum für Naturkunde an der Humboldt-Universität Berlin; Institut Royal des Sciences Naturelles de Belgique, Brussels; the Australian Museum, Sydney; the South Australian Museum, Adelaide; Museum of Comparative Zoology, Harvard, Cambridge; American Museum of Natural History, New York; and the United States National Museum, Washington, D. C. Eguchi's [1965] specimens could not be found at the National Institute of Polar Research, Tokyo, or the Tokyo Kasei University.

Synonymies are complete unless otherwise indicated.

In the sections on material the numbers in parentheses indicate the number of specimens in each lot. The number of specimens is not indicated for colonial species. Following this information, or the station number for colonial species, is an indication of where the specimen is deposited. If the vessel's cruise number is included, it precedes the station number and is linked to it with a hyphen.

Most of the holotypes and paratypes of the new species are deposited at the United States National Museum; the remainder are at the British Museum.

A confirmed depth range is used to avoid erroneous end points resulting from bathymetrically wide-ranging trawls. The stated bathymetric range for each species extends from the deepest shallow to the shallowest deep component of the trawled depth ranges [see Cairns, 1979, p. 10].

1

TABLE 1. Chronological Listing of Research on Antarctic and Subantarctic Scleractinia

Reference	Vessel and/or Expedition	Remarks
Milne Edwards and Haime [1848]	Du Petit-Thouars, Venus	3 Specimens of Flabellum thouarsii from off Falkland Islands; deposited at MNHNP.
Moseley [1876]	HMS Challenger	Preliminary report on Challenger Scleractinia, including records of 3 species from off Tristan da Cunha Group and Prince Edward Islands; deposited at BM.
Studer [1878]	SMS Gazelle	2 specimens of Flabellum thouarsii reported from off southern Argentina; specimens deposited at Museum für Naturkunde, Berlin.
Moseley [1881]	HMS Challenger	Final report on Challenger Scleractinia, including 10 species (5 of them new) from various Subantarctic localities, including off eastern and western southern South America, off Tristan da Cunha Group, and off Prince Edward Islands; deposited at BM.
von Marenzeller [1903]	SY Belgica, Expédition Antarctique Belge (1897-1899)	5 specimens of 2 species from 3 stations off Peter I Island, Antarctica; deposited at Institut Royal des Sciences Naturelles de Belgique, Brussels.
von Marenzeller [1904a]	Valdivia, Deutsche Tiefsee-Expedition (1898-1899)	2 specimens of C. antarctica from 1 station off Bouvetøya; deposited at Museum für Naturkunde, Berlin.
Pax [1910]	Gauss, Deutsche Südpolar-Expedition (1901-1903)	7 specimens of 3 species from 5 stations off Gaussland, Antarctica; histology of Flabellum impensum.
Gardiner [1913]	Scotia, Scottish National Antarctic Expedition (1902-1904)	3 specimens of Caryophyllia profunda from 1 station off Gough Island.
Gravier [1914a]	Pourquoi-Pas? Deuxième Expédition Antarctique Française (1908-1910)	3 specimens of Desmophyllum antarcticum (< Javania antarctica) from 2 stations off Palmer Archipelago; deposited at MNHNP.
Gravier [1914b]	Pourquoi-Pas? Deuxième Expédition Antarctique Française (1908-1910)	13 specimens of 4 species from 3 stations off Antarctic Peninsula; deposited at MNHNP.
David and Priestley [1914]	Nimrod British Antarctic Expedition (1907-1909)	Fossil Gardineria antarctica from Mount Larsen; deposited at the Australian Museum, Sydney.
Gardiner [1929a]	Terra Nova, British Antarctic Expedition (1910)	19 specimens of 3 species from 4 stations in Ross Sea; deposited at BM.
Thomson and Rennet [1931]	Aurora Australasian Antarctic Expedition (1911-1914)	7+ specimens of 3 species from 2 stations off Queen Mary Coast, Antarctica. Another 2 scleractinian species from off Maria Island, Tasmania; all but F. australe are misidentified; deposited at the Australian Museum, Sydney.
Gardiner [1939]	RRS Discovery, RSS William Scoresby (1926-1933)	280 specimens of 12 nominal species from 33 stations off southern South America and Antarctic Peninsula; 9 of 12 species reassigned; deposited at BM.
Niino [1958]	Umitaka-Maru, JARE	4 specimens of 4 species from 2 stations off Riiser-Larsen Peninsula (Cape Cook), Antarctica; all species misidentified; deposition unknown.

TABLE 1. (continued)

Reference	Vessel and/or Expedition	Remarks
Wells [1958]	*Discovery*, BANZARE (1929-1931)	97 specimens of 5 species from 10 stations off coast of Antarctica between 40° and 80°E and 110° and 120°E; 3 of 5 species reassigned; deposited at the South Australian Museum, Adelaide.
Squires [1961]	R/V *Vema* (cruise 14)	92 specimens of 4 species (including 2 new species) from 6 stations off the Falkland Islands and Tierra del Fuego; desposited at AMNH.
Speden [1962]		Review of fossil records of *Gardineria antarctica* on Antarctic continent.
Squires [1962b]	HMNZS *Endeavour* (1958-1960)	242 specimens of 5 species (including 2 new species) from 13 stations in the Ross Sea; reviews distribution of all Antarctic Scleractinia; presumably deposited at NZOI.
Squires [1963a]	United States Exploring Expedition (1838-1842)	2 fossilized specimens of 2 species collected at Tierra del Fuego; deposited at MCZ.
Squires [1964c]		Discussion of research potential of Antarctic Scleractinia; no specimens reported.
Squires [1965b]	HMNZS *Endeavour* (Macquarie Gap Cruise)	Specimens of 3 species from 1 station east of the Auckland Islands; deposited at NZOI and USNM.
Eguchi [1965]	*Umitaka-Maru* and *Soya*, JARE	9 specimens of 9 nominal species (including 2 new species) from 6 stations off Riiser-Larsen Peninsula (Cape Cook), Antarctica; specimens overlap with those of Niino [1958]; less than half of species correctly identified; deposition unknown.
Bullivant [1967]	HMNZS *Endeavour*	Description of *Gardineria antarctica* assemblage in Ross Sea; specimens previously reported by Squires [1962b].
Squires [1969]		Reviews distribution of all Antarctic Scleractinia; provides distribution maps for 23 species; includes new records but does not document them (presumably based on USARP and NZOI specimens); deposited at USNM and probably NZOI.
Keller [1974]	D/E *Ob*, *Academic Kurchatov*	357 specimens of 2 species from 3 stations from off southeastern South America and Falkland Islands; *F. antarcticum* misidentified; deposition unknown.
Podoff [1976]	miscellaneous	Examination of microstructure by SEM of 6 Antarctic species; unpublished M. A. thesis; deposited at USNM.
Sorauf and Podoff [1977]	miscellaneous	Examination of microstructure of 3 Antarctic species; deposited at USNM.
Present study	USNS *Eltanin*; ARA *Islas Orcadas*; R/V *Hero* (USARP program); USS *Atka*; USS *Burton Island*; USS *Staten Island*; USS *Edisto* (Deep Freeze expeditions); HMNZS *Endeavour*, *Rotoiti*, *Viti* (NZOI vessels); R/V *Vema*, *Walther Herwig*; USCGC *Eastwind*, cruise 66	8700 specimens of 39 species (including 7 new species) from 310 stations throughout Antarctic and Sub-antarctic; primarily deposited at USNM, also at AMNH and MCZ.

Some specimens have been coated with dark dye and recoated with a fine layer of ammonium chloride in order to improve their contrast for photography. These specimens are noted in the plate legends.

The following abbreviations are used in the text:

Vessels

EAD	USCGC _Eastwind_.
EW	USCGC _Eastwind_, cruise 66.
GLD	USS _Glacier_, Deep Freeze IV Expedition.
NZOI	collected by the New Zealand Oceanographic Institute, including the HMNZS _Endeavour_, _Rotoiti_, and _Viti_.
PD	pebble dredge (used in conjunction with some R/V _Vema_ stations).
USARP	United States Antarctic Research Program.
WH	_Walther Herwig_.
WS	RSS _William Scoresby_.

Museums

AMNH	American Museum of Natural History, New York.
BM	British Museum (Natural History), London.
MCZ	Museum of Comparative Zoology, Harvard, Cambridge.
NMNH	see USNM.
MNHNP	Muséum National d'Histoire Naturelle, Paris.
NZOI	New Zealand Oceanographic Institute, Wellington.
SME	Station Marine d'Endoume, Marseille (most of these specimens will be deposited at the Muséum National d'Histoire Naturelle, Paris).
USNM	United States National Museum, Smithsonian Institution, Washington, D. C.
ZIZM	Zoologisches Institut und Zoologisches Museum, Hamburg.
ZMA	Zoölogische Museum, Amsterdam.

Other

GCD	greater calicular diameter.
LCD	lesser calicular diameter.
CD	calicular diameter.
PD	pedicel diameter.
H	height.
S_x, C_x, P_x, CS_x	septa, costae, pali, or costosepta of cycle designated by numerical subscript.
SEM	scanning electron microscopy.

Checklist of Species Known From the Antarctic and Subantarctic Regions

Order SCLERACTINIA Bourne, 1900
Suborder FUNGIINA Verrill, 1865
Family FUNGIIDAE Dana, 1846
Fungiacyathus Sars, 1872

F. marenzelleri (Vaughan, 1906)
F. fragilis G. O. Sars, 1872

Family MICRABACIIDAE Vaughan, 1905
Leptopenus Moseley, 1881

L. sp. cf. L. discus Moseley, 1881

Suborder FAVIINA Vaughan and Wells, 1943
Superfamily FAVIICAE Gregory, 1900
Family RHIZANGIIDAE d'Orbigny, 1851
Astrangia Milne Edwards and Haime, 1848

A. rathbuni Vaughan, 1906

Phyllangia Milne Edwards and Haime, 1848

P. fuegoensis Squires, 1963

Family OCULINIDAE Gray, 1847
Bathelia Moseley, 1881

B. candida Moseley, 1881

Madrepora Linnaeus, 1758

M. oculata Linnaeus, 1758

Suborder CARYOPHYLLIINA Vaughan and Wells, 1943
Superfamily CARYOPHYLLIICAE Gray, 1847
Family CARYOPHYLLIIDAE Gray, 1847
Subfamily CARYOPHYLLIINAE Gray, 1847
Caryophyllia Lamarck, 1801

C. antarctica Marenzeller, 1904
C. squiresi n. sp.
C. profunda Moseley, 1881
C. eltaninae n. sp.
C. mabahithi Gardiner and Waugh, 1938

Cyathoceras Moseley, 1881

C. irregularis n. sp.

Aulocyathus Marenzeller, 1904

A. recidivus (Dennant, 1906) n. comb.

Subfamily TURBINOLIINAE Milne Edwards and Haime, 1848

Sphenotrochus Milne Edwards and Haime, 1848

S. gardineri Squires, 1961

Subfamily DESMOPHYLLIINAE Vaughan and Wells, 1943
Desmophyllum Ehrenberg, 1834

D. cristagalli Milne Edwards and Haime, 1848
forma cristagalli Milne Edwards and Haime, 1848
forma ingens Moseley, 1881
forma capense Gardiner, 1904

Lophelia Milne Edwards and Haime, 1849

L. prolifera (Pallas, 1766)

Subfamily PARASMILIINAE Vaughan and Wells, 1943
Solenosmilia Duncan, 1873

S. variabilis Duncan, 1873

Goniocorella Yabe and Eguchi, 1932

G. dumosa (Alcock, 1902)

Superfamily FLABELLICAE Bourne, 1905
Family FLABELLIDAE Bourne, 1905
Flabellum Lesson, 1831

F. thouarsii Milne Edwards and Haime, 1848
F. areum n. sp.
F. curvatum Moseley, 1881
F. impensum Squires, 1962
F. flexuosum n. sp.
F. gardineri n. sp.
F. knoxi Ralph and Squires, 1962
F. apertum Moseley, 1876
  forma apertum Moseley, 1876
  forma patagonichum Moseley, 1881
F. truncum n. sp.

Javania Duncan, 1876

J. cailleti (Duchassaing and Michelotti, 1864)
J. antarctica (Gravier, 1914) n. comb.

Gardineria Vaughan, 1907

G. antarctica Gardineria, 1929

Family GUYNIIDAE Hickson, 1910
Stenocyathus Pourtalès, 1871

S. vermiformis (Pourtalès, 1868)

Suborder DENDROPHYLLIINA, Vaughan and Wells, 1943
Family DENDROPHYLLIIDAE Gray, 1847
Balanophyllia Wood, 1844

B. malouinensis Squires, 1961
B. sp.
B. chnous Squires, 1962

Enallopsammia Michelotti, 1871

E. rostrata (Pourtalès, 1878)
E. sp. cf. E. marenzelleri Zibrowius, 1973

Uncertain Records

Caryophyllia clavus var. smithi sensu Moseley, 1881
Flabellum transversale conicum sensu Eguchi, 1965
Flabellum ongulense Eguchi, 1965
Desmophyllum pseudoseptatum Eguchi, 1965

Species Account

Order SCLERACTINIA Bourne, 1900
Suborder FUNGIINA Verrill, 1865
Family FUNGIIDAE Dana, 1846
Genus Fungiacyathus Sars, 1872

Diagnosis. Solitary, cupolate, free. Septotheca thin; costae thin and spinose. Septa irregularly dentate, laterally braced by thin ribbons extending from septotheca and by thin septal striae. Columella feeble. Paliform lobes sometimes present. Type-species: Fungiacyathus fragilis Sars, 1872, by monotypy.

1. Fungiacyathus marenzelleri (Vaughan, 1906)
Plate 1, figs. 1, 2, 8

Bathyactis symmetrica; Moseley, 1881, pp. 186-190 (part: Challenger sta. 133, 147, 157, 299, 325, 332), pl. 11, figs. 1-5, 7.--von Marenzeller, 1904b, p. 76.--Gravier, 1920, p. 97 (part).--? Eguchi, 1965, pp. 5-7, pl. 1, figs. 4a-4c.
Fungia symmetrica; Duncan, 1873, p. 334, pl. 49, figs. 16-19.
Bathyactis marenzelleri Vaughan, 1906b, p. 66, pl. 4, figs. 1, 1a, 1b.
Fungiacyathus symmetricus; Wells, 1958, pp. 267, pl. 2, figs. 1, 2.--Squires, 1962b, p. 13; 1969, p. 17, pl. 6, map 2.
Fungiacyathus marenzelleri; pl. 7, figs. A-K.-- Cairns, 1979, pp. 35-37, pl. 2, figs. 8, 9, pl. 3, figs. 3, 8. Zibrowius, 1980, pp. 24, 25, pl. 6, figs. A-M.

Description. Base of corallum flat, center of base slightly raised. Largest specimen known measuring 40 mm across base; largest Antarctic specimen 38.5 mm in diameter. Forty-eight thin, ridged costae radiating from center of base, becoming more raised and sinuous toward calicular edge. Costae sometimes projecting as much as 1.5 mm from base at calicular edge. Base extremely thin and fragile, sometimes perforate, especially toward calicular edge. Five to seven synapticular plates occurring in every interseptal space, becoming increasingly larger and more oblique toward edge of corallum.

Septa hexamerally arranged in four cycles. S_1 largest septa and only ones reaching columella without fusion to other septa. Septa of remaining cycles progressively smaller. Each S_2 reaching columella, there joined by pair of S_3, all loosely fused or covered over by triangular canopy composed of thin calcareous plate. Pairs of S_4 fused to S_3 by similar but larger and higher canopies, these extending from S_2-S_3 canopies to halfway to calicular edge. These canopies often perforate.

Septa laterally carinate, S_1 possessing about 7-10 carinae, or about 1 every 1.9 mm. Carinae about 0.4 mm high and usually symmetrical on both sides of septum. Near columella, carinae more closely spaced and corresponding to high septal spines. Carinae vertical near columella, oblique midway between columella and calicular edge and almost horizontal near calicular edge. Most carinae extending from septal edge to base; some shorter, extending only halfway to base; some branching from other carinae. Near base most carinae curving toward columella, often degenerating into rows of granules. If still solid, carinae may be confused with synapticulae but can usually be distinguished by their more oblique orientation, often intersecting synapticulae at an acute angle. Septa bearing elongate, slender spines near columella but becoming less serrate and usually lobate in profile toward calicular edge. Height of exsert lobes up to 10 mm above base. All septa but S_4 bearing lobes, these lobes damaged in most specimens. Septa extraordinarily fragile and specimens rarely collected fully intact. Septal edges straight to irregularly sinuous. Columella variable in size, composed of loose fusion of inner septal spines and additional trabeculae.

Discussion. F. marenzelleri was frequently reported as Bathyactis or F. symmetricus by earlier authors, probably because of Moseley's [1881] assumption that all small specimens were simply juveniles of a species with a larger adult corallum. F. symmetricus has subsequently been shown to be endemic to the western Atlantic (183-1644 m) and to be rarely larger than 14 mm in CD [Cairns,

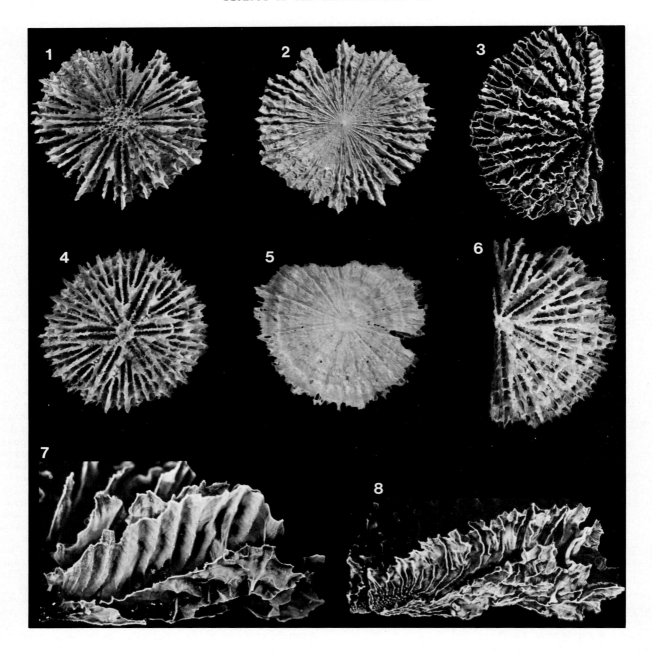

Plate 1. Fungiacycathus

1, 2, 8. Fungiacyathus marenzelleri (Vaughan): 1, 2, USNM 47476, Eltanin sta. 134,
          CD = 32.3 mm; 8, USNM 47477, Eltanin sta. 1545, x4.3, several septa coated
          with ammonium chloride.
   3-7.   Fungiacyathus fragilis Sars: 3, 6, 7, USNM 47536, Eltanin sta. 1412, CD =
          24.4 mm (3 and 7 coated with ammonium chloride); 4, holotype of F.
          hawaiiensis Vaughan, USNM 20834, Albatross sta. 4125, CD = 23.1 mm; 5, USNM
          47537, Eltanin sta. 1846, CD = 21.3 mm, base.

1979]. F. marenzelleri is distinguished from F. symmetricus by its much larger corallum, broader distribution, and greater depth range. F. symmetricus also has higher septal spines and a more solid, well-defined base.

Specimens from four localities (Eltanin stations 138, 426, 993, and 1545) differ from typical F. marenzelleri by having a much more crowded arrangement of septal carinae (about 20 per septum, or 1 every 0.8 mm) and more spinose septal faces. These specimens, all from the South Shetland Islands and east of South Orkney Islands, may represent a new species, but until the variation of F. marenzelleri is better understood, they are assigned to this species. Eguchi's [1965] B. symmetrica also seems to belong to this form.

When the deeper records of Fungiacyathus spp. are reexamined [e.g., Moseley, 1881; Gardiner and Waugh, 1939; Keller, 1976], more synonyms for F. marenzelleri may result, and its geographic range may thus be increased. All records deeper than 1800 m should be reevaluated for this possibility. This species may eventually be found to be the most widespread and deepest-living species of scleractinian coral.

Material. Eltanin sta. 13 (5), USNM 47470; sta. 18 (23), USNM 47466; sta. 20-134 (5), USNM 47476; sta. 138 (3), USNM 47475; sta. 353 (1), USNM 47471; sta. 426 (8), USNM 45673; sta. 993 (1), USNM 47473; sta. 997 (6), USNM 47474; sta. 1148 (1), USNM 47469; sta. 1545 (3), USNM 47477; sta. 1957 (1), USNM 47466. Hero sta. 721-1081 (4), USNM 47468. Glacier sta. 11 (6), USNM 47465. Albatross sta. 4397 (1), USNM 47467. Specimen identified as F. symmetricus by Wells [1958], South Australian Museum H 70; specimens listed by Cairns [1979], USNM. Types of B. marenzelleri.

Types. The holotype of F. marenzelleri is deposited at the United States National Museum (47415); three paratypes are at the Museum of Comparative Zoology. Type-locality: 8°07.5'S, 104°10.5'W (off Peru); 3820 m.

Distribution. Widely distributed throughout Atlantic Ocean as far north as Greenland; eastern Pacific; circum-Subantarctic (including off Îles Crozet); off South Shetland Islands; east of South Orkney Islands; Weddell Sea; ? off Prince Harald Coast and Enderby Land, Antarctica (Map 1). Depth range: 300-5870 m.

There is a direct relationship between depth of occurrence and proximity to the Antarctic, the more southerly records being shallower. The shallowest records of this species (300-500 m) are represented by the four continental Antarctic records; the four records from the South Shetland Islands range from 300 to 1435 m. No other record is shallower than 1805 m.

2. Fungiacyathus fragilis G. O. Sars, 1872
Plate 1, figs. 3-7

Fungiacyathus fragilis M. Sars, 1869, pp. 250, 265, 274 (nomen nudum).--G. O. Sars, 1872; p. 58, pl. 5, figs. 24-32. Cairns, 1979, p. 206. --Zibrowius, 1980, pp. 23, 24, pl. 5, figs. A-J.
Bathyactis symmetrica; Verrill, 1882, p. 313; 1883, p. 65.--Gravier, 1920, p. 97 (part).--Thomson, 1931, p. 9.
Bathyactis hawaiiensis Vaughan, 1907, pp. 145, 146, pl. 27, figs. 1, 1a.

Fungiocyathus fragilis; Jungersen, 1916, p. 35 (part).--Broch, 1927, p. 8.--Nordgaard, 1929, p. 103.

Description. Base of corallum flat, very thin, sometimes irregularly perforate. Evidence of regeneration from fragments of corallum. Largest specimen reported 45 mm in basal diameter; largest specimen from Subantarctic 25 mm in diameter. Slightly ridged C_{1-3} radiate from center of base.

Septa hexamerally arranged in five complete cycles, fifth cycle appearing at CD of 9-10 mm. Septal arrangement similar to previously described species: S_1 independent; inner edges of septa of remaining cycles fused to one another by thin, perforate triangular lamellae (canopies). Each larger septum forming large nonserrate lobe for most of its length, with few, if any, projecting spines near columella. Height of septal lobes up to one fifth to one fourth of CD. These lobes, as well as all septa, with highly sinuous outer and upper edges, corresponding to septal corrugations. Corrugations vertical near upper edge, becoming more horizontal as they curve toward columella near base. About 12 corrugations per septum, or 1/mm, giving septa 'wrinkled' aspect. Crests of corrugations regularly spaced, usually rounded and smooth, but may bear row of low, pointed granules or may even be slightly carinate. Small, pointed granules usually on all septa near columella. Septa extraordinarily fragile. All septa joined to adjacent septa by synapticular plates, these plates increasing in size toward calicular edge. About 7-10 plates occurring per centimeter, continuing to add on as corallum increases in diameter.

Columella round and small, sometimes a solid, horizontal lamella but usually a loose fusion of inner edges of larger septa.

Discussion. There are four other nominal Recent species of Fungiacyathus with five cycles of septa. F. fragilis differs from western Atlantic F. pusillus (Pourtalès, 1868) in being larger and having sinuous septa, from Indian Ocean F. stephanus (Alcock, 1893) in having a flat base and lower septa, and from Indian Ocean and eastern Pacific F. paliferus (Alcock, 1902) in lacking paliform lobes. It is indistinguishable from F. hawaiiensis (Vaughan, 1907); however, comparisons involving only one specimen (the holotype) cannot be conclusive. F. kikaiensis (Yabe and Eguchi, 1942), also with five cycles, is from the Pliocene-Pleistocene of Japan.

Material. Eltanin sta. 1412 (5+), USNM 47536; sta. 1846 (3), USNM 47537. Specimens of Verrill [1882], (Yale Peabody Museum, New Haven) YPM 8322 and USNM 47538-47539; specimens of Zibrowius [1980], SME. Holotype of B. hawaiiensis.

Types. One syntype of F. fragilis is deposited at the Oslo Museum (B626). Type-locality: 'Skraaven in Lofoten'; 549 m. The holotype of B. hawaiiensis is deposited at the United States National Museum (20834). Type-locality: between Oahu and Kauai islands, Hawaii; 1761-2056 m.

Distribution. Eastern Atlantic in area bordered by Norway, Cape Verde Islands, and the Azores; off Massachusetts; off Hawaii; west of South Island, New Zealand; Macquarie Ridge (Map 1). Worldwide depth range: 285-2200 m; New Zealand-Macquarie records: 1647-1693 m.

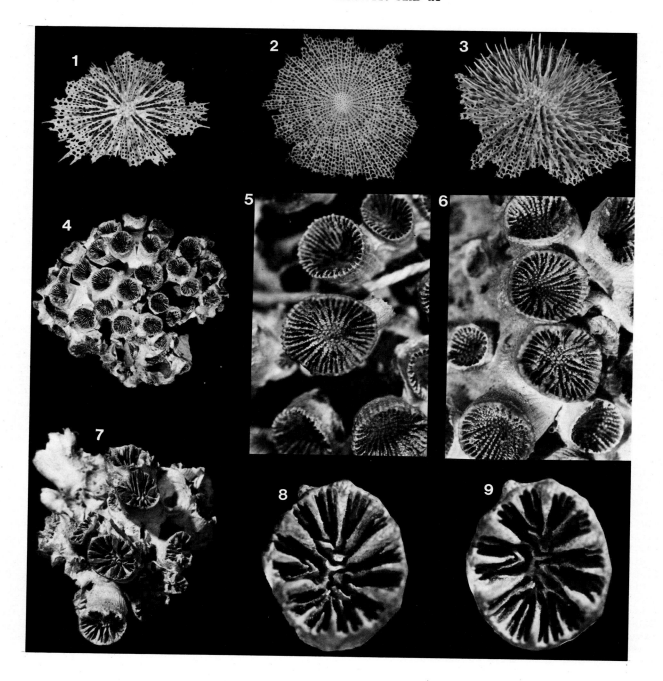

Plate 2. _Leptopenus_, _Astrangia_, and _Phyllangia_

1-3. _Leptopenus_ sp. cf. _L. discus_ Moseley: 1, USNM 47481, _Eltanin_ sta. 1545, CD = 18.2 mm; 2, USNM 47483, _Eltanin_ sta. 2002, CD = 16.5 mm, base; 3, same specimen, calice.

4-6. _Astrangia rathbuni_ Vaughan: 4, 6, specimen reported by Squires [1963a], MCZ 2520, CD = 5-6 mm; 5, holotype, USNM 10974, CD = 6-7 mm.

7-9. _Phyllangia fuegoensis_ Squires: holotypic colony and calices, MCZ 5390, CD = 7.5 x 6.5 mm.

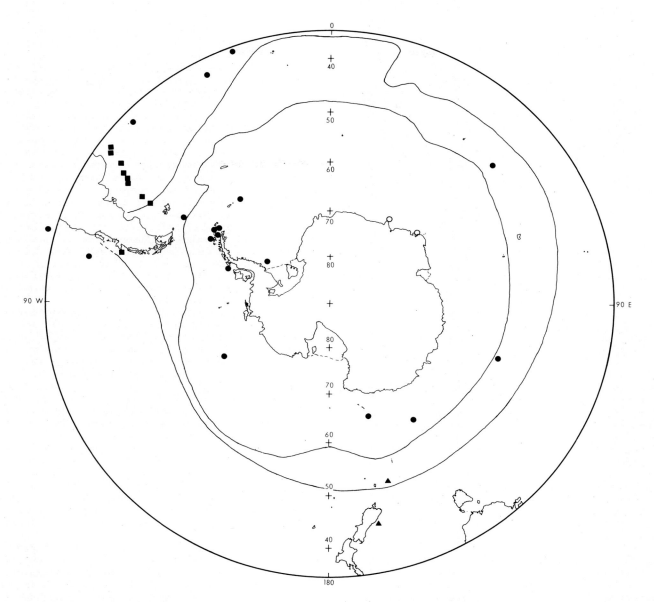

Map 1. Distribution of <u>Fungiacyathus</u> <u>marenzelleri</u> (solid circles), <u>Fungiacyathus</u> <u>fragilis</u> (solid triangles), and <u>Bathelia</u> <u>candida</u> (solid squares).

Family MICRABACIIDAE Vaughan, 1905
Genus <u>Leptopenus</u> Moseley, 1881

<u>Diagnosis</u>. Solitary, discoidal, free. No wall, costae alternating in position with septa. Costae and septa united by simple synapticulae producing very porous, delicate corallum. Columella trabecular. Type-species: <u>Leptopenus</u> <u>discus</u> Moseley, 1881, by subsequent designation [Wells, 1936].

3. <u>Leptopenus</u> sp. cf. <u>L</u>. <u>discus</u> Moseley, 1881
Plate 2, figs. 1-3

<u>Leptopenus</u> <u>discus</u> Moseley, 1881, pp. 205-208, pl. 14, figs. 1-4, pl. 16, figs. 1-7.--Not <u>L</u>. <u>discus</u> Dennant, 1906, p. 162 (? <u>Letepsammia</u> sp.).--Wells, 1958, p. 262.--Squires, 1965a, pp. 878, 879, fig. 1; 1967, p. 505; 1969, p. 17, pl. 6, map 2.--Keller, 1977, p. 37, fig. 1.--Cairns, 1979, pp. 37, 38, pl. 3, figs. 4-7.

<u>Leptonemus</u> <u>discus</u>; Agassiz, 1888, p. 154, fig. 479 (misspelling).

<u>Description</u>. Corallum discoidal and extremely fragile. Base of corallum flat to slightly concave, especially near center. Largest Antarctic specimen 18.0 mm in CD.

Septa hexamerally arranged in two cycles and multiple bifurcation of septa of third cycle, unlike species from any other family of Scleractinia. (The traditional method of designating septa of higher cycles as pairs of new septa flanking each previously formed septum does not apply to <u>Leptopenus</u>, and therefore a new system of terminology is introduced here (Text figure 1).) Each S_3 bifurcating about 2 mm from columella, the two resulting septa being referred to as S_3'. Not far from this junction the S_3' bifurcate, each forming two S_3''. Sometimes, near edge of calice, pair of S_3''' forming, as described by Moseley for syntypes. Branching of S_3 not always symme-

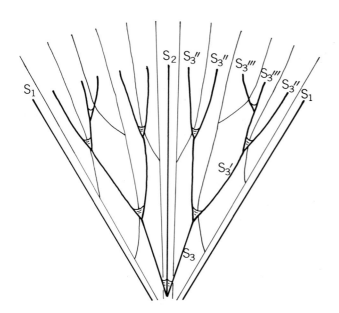

Fig. 1. Diagrammatic representation of one system of *Leptopenus* *discus*. Heavy lines represent septa, light lines costae.

trical within one system or consistent in systems of one corallum. (For instance, in a half system of a specimen from Eltanin station 2002, one S_3' produced only two S_3'', whereas the other S_3' produced two S_3'' and four S_3'''. Also, one of the syntypes [Moseley, 1881, pl. 14, fig. 1] is illustrated as having a pair of S_3'''' in the upper right system. The regular arrangement of 72 septa with one pair of S_3''' per system to which Moseley alluded is probably just one variation of septal arrangement.)

S_1 only independent septa, extending to columella and bearing eight or nine long, recurved spines. S_2 also extending to columella and bearing five or six similar spines, with pair of S_3 joining each S_2 close to columella S_3 and all of its bifurcations bearing similar recurved spines decreasing in size toward calicular edge. Thin, triangular canopy usually at point of junction of S_3 to S_2 and at every bifurcation of S_3. Septal spines of S_2 just distal of S_2-S_3 junction usually tallest spines on corallum, rising about 1 mm above columellar spines. Septa very low and solid, composed mainly of large spines united by thin lamella. No septal granulation. Columella a very spiny mound in center of calice, merging with inner septal spines.

Base of corallum consisting of radiating network of bifurcating costae, these costae alternating in position with overlaying septa. Costae not arranged like septa, because all costae bifurcate; none independent (Text figure 1). Each costa attached to its two adjacent septa by thin synapticular bridges; space between bridges forming elliptical pores. Viewed from above or below, corallum appearing regularly perforate. Pores increasing in size from center to calicular edge. For inner two thirds of corallum, synapticular bridges reaching upward from costae to meet septa; in outer part of corallum, however, septa becoming rudimentary or absent and synapticular bridges becoming wider and horizontal, but pores still

persist. At calicular edge, costae forming thin spines projecting up to 1.6 mm beyond synapticular zone, or 17-20% of calicular radius.

Discussion. The Antarctic specimens are only provisionally identified as L. discus because (1) they are taller than the syntypes, (2) the septal arrangement is slightly different, and (3) they sometimes lack canopies at the junctions of septa. The reasons that these specimens are taller (up to 3.8 mm) may be the slight concavity of the base or their intact septal spines. Most of the spines are broken off of the syntypes, which measure less than 2 mm in height. The septal arrangement and presence or degree of development of canopies may be a matter of individual variation. Variation in this species is very poorly known. Only 6 specimens of L. discus have been reported previously, 2 of which were fragments. Although the 44 specimens reported herein were all collected with tissue intact, upon cleaning, much of the corallum fell apart, especially the peripheral areas. Additional whole specimens are needed to study variation of septal arrangement and costal spines.

L. discus is distinguished from L. irinae Keller, 1977, by its shorter costal and peripheral septal spines. It is easily distinguished from L. hypocoelus Moseley, 1881, by its lesser height and smaller S_2 spines and from L. solidus Keller, 1977, by its alternating septa and costae.

Material. Eltanin sta. 598 (1), USNM 47480; sta. 1545 (2), USNM 47481; sta. 1926 (1), USNM 47479; sta. 2002 (3), USNM 47483; sta. 2108 (37), USNM 47482. Specimen of Agassiz [1888], USNM 46916. Syntype from Challenger sta. 147.

Types. The four syntypes of L. discus, collected at Challenger stations 147, 157, and 323, are deposited at the British Museum. The specimen from Challenger station 147 is numbered 1880.11.25.159. Type-locality: since a lectotype has not been designated, the type-localities are off Río de la Plata, South America, and Subantarctic Indian Ocean; 2926-3566 m.

Distribution. Off Cuba; off Río de la Plata; Makassar Strait, Indonesia; off South Orkney Islands; off South Sandwich Islands; Subantarctic Indian Ocean (including off Îles Crozet); Ross Sea (Map 2). Worldwide depth range: 2000-3566 m; Antarctic records: 2035-2384 m.

Suborder FAVIINA Vaughan and Wells, 1943
Superfamily FAVIICAE Gregory, 1900
Family RHIZANGIIDAE d'Orbigny, 1851
Genus Astrangia Milne Edwards and Haime, 1848

Diagnosis. Colonial, extratentacular budding forming cerioid, plocoid, or reptoid coralla. Corallites united basally by thin coenosteum or stolons. Septa dentate; columella papillose. Type-species: Astrangia michelini Milne Edwards and Haime, 1848, by subsequent designation [Milne Edwards and Haime, 1850].

4. Astrangia rathbuni Vaughan, 1906
Plate 2, figs. 4-6

? Astrangia Verrill, 1869, p. 526.
Astrangia rathbuni Vaughan, 1906a, pp. 849-850, p. 78, figs. 1-3.--Squires, 1963a, pp. 10, 11, figs. 3-7.--Laborel, 1971, pp. 200, 201, pl. 6, fig. 1, map 7.--Zibrowius, 1974c, pp. 165, 166.--Not A.

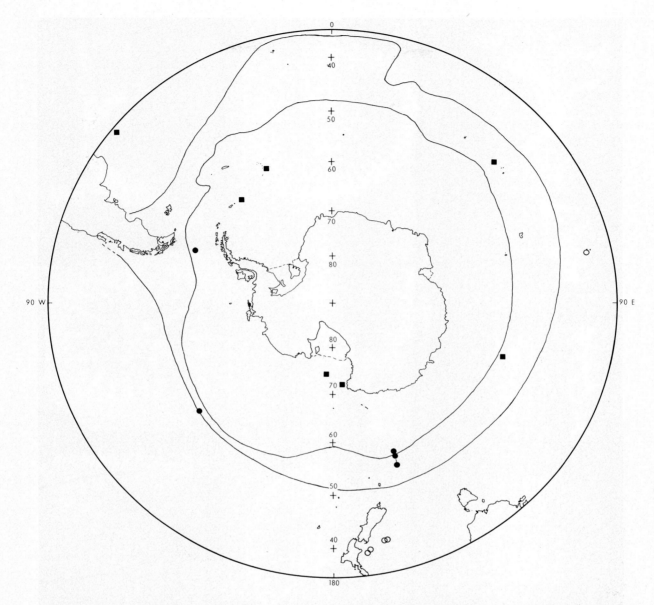

Map 2. Distribution of *Madrepora oculata* (solid circles), *Madrepora vitiae* (open circles), and *Leptopenus discus* (solid squares).

rathbuni; Avent et al., 1977, p. 200 (is *A. astreiformis* Milne Edwards and Haime, 1849).
? *Astrangia* sp. Pax, 1910, p. 74.--Gravier, 1914b, p. 121.--Squires, 1961, p. 20.

Description. Small encrusting colonies 30-50 mm in diameter; corallites united by basal coenosteum. Cylindrical corallites 5-6 mm in diameter projecting up to 9 mm from coenosteum. Sometimes extratentacular budding from other corallites. Corallites weakly costate, usually brownish.

Septa hexamerally arranged in four systems. S_1, S_2, and S_3 equal in size, sloping gradually toward deep fossa. S_4 half as large and joining adjacent S_3 halfway to columella. Inner edges of all septa highly dentate (beaded), each bearing 7-10 irregular teeth. Columella indistinguishable from lower septal edges, consisting of mass of similarly shaped teeth.

Remarks. An X ray diffraction analysis of a small fragment from specimen MCZ 2520 revealed a very high calcite peak, indicating a subfossil age for the specimen.

Discussion. Verrill [1869], in a general discussion of the genus *Astrangia*, mentioned that one species (not named) occurred in the Strait of Magellan. Pax [1910], Gravier [1914b], and Squires [1961] perpetuated this obscure record as *Astrangia* sp., none of whom documented its occurrence. Zibrowius [1974c] suggested that both Verrill's [1869] report of the unnamed *Astrangia* and Squires's [1963a] record of *A. rathbuni* from off Tierra del Fuego were based on the same specimens. Both authors were familiar with the Museum of Comparative Zoology coral collection, where Squires's specimens are deposited. Despite their fossilized condition, Squires's [1963a] Tierra del Fuego specimens are definitely *A. rathbuni*.

Material. Specimens of Squires [1963a], MCZ 2520; specimens (*A. astreiformis*) of Avent et al.

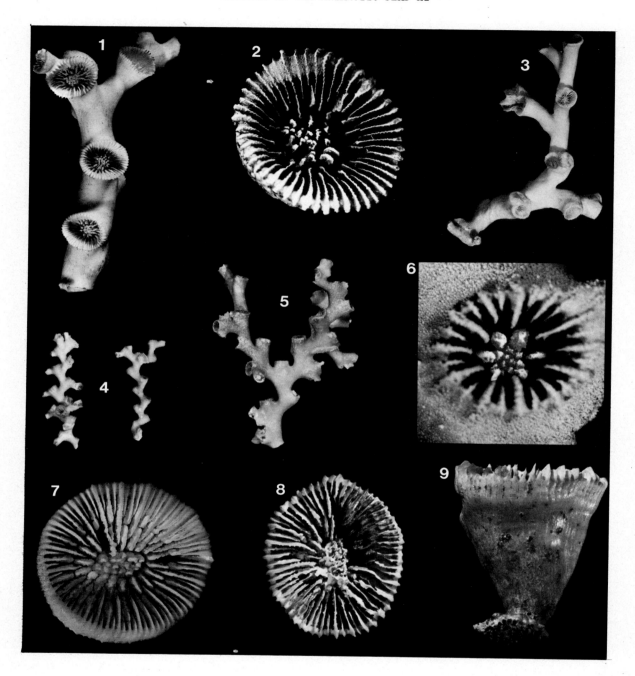

Plate 3. *Bathelia*, *Madrepora*, and *Caryophyllia*

1-3.   *Bathelia* *candida* Mosely: 1, syntype branch, BM, *Challenger* sta. 320, x1.2;
       2, USNM 47512, *Vema* sta. 17-14, GCD = 9.5 mm; 3, same specimen, H = 98 mm.
4-6.   *Madrepora* *oculata* Linnaeus: 4, USNM 47499, *Eltanin* sta. 1346, x0.8; 5,
       specimen from same lot, x1.3; 6, USNM 47515, NZOI sta. B-314, CD = 2.1 mm,
       coated with ammonium chloride, forma *vitiae*.
7-9.   *Caryophyllia* *antarctica* Marenzeller: 7, USNM 45677, *Eltanin* sta. 138, GCD =
       28.7 mm; 8, 9, syntype, Museum für Naturkunde 5067, *Valdivia* sta. 127, GCD
       = 20.0 mm, H = 21.6 mm.

[1977], Florida Department of Natural Resources FSBC I 14492. Holotype.

Types. The holotypic colony of A. rathbuni is deposited at the United States National Museum (10974). Type-locality: Paqueta, Rio de Janeiro.

Distribution. Off eastern coast of South America from 22°S to 37°59'S; Tierra del Fuego fossil specimens collected by United States Exploring Expedition (1838-1842) but unfortunately without precise locality data. Several to 90 m.

Genus Phyllangia Milne Edwards and Haime, 1848

Diagnosis. Colonial, extratentacular budding forming reptoid colonies, united basally by thick coenosteum. Inner septal edges smooth to slightly dentate. Columella rudimentary; P_3 sometimes present. Type-species: Phyllangia americana Milne Edwards and Haime, 1849, by subsequent designation [Milne Edwards and Haime, 1850].

5. Phyllangia fuegoensis Squires, 1963
Plate 2, figs. 7-9

Phyllangia fuegoensis Squires, 1963a, pp. 13, 16, figs. 1, 2.

Description. In the following the holotypic colony is described. Fragmentary corallum very worn, perhaps the result of fossilization. Fragment about 35 x 20 x 20 mm, bearing 21 corallites. Corallites cylindrical, encrusting a bivalve shell and budding from parent corallites. Corallites up to 10-15 mm tall with elliptical calices measuring up to 7.5 x 6.5 mm in diameter. Septa hexamerally arranged in four cycles. S_1 thickest and most exsert; septa of remaining cycles progressively smaller. Inner septal edges worn but appear to be straight, entire, and vertical, except for S_3, these usually having small, pointed paliform lobes. Pair of lobes often meeting or at least bending toward each other across each S_2. Low, pointed granules on septal faces. Columella rudimentary, crispate, surrounded by paliform lobes.

Discussion. Like the preceding species, P. fuegoensis was also collected by the United States Exploring Expedition from Tierra del Fuego, without precise locality. Contrary to Squires's [1963a] comparison, it is quite distinct from Astrangia floridana (Gane, 1895) but very similar to P. americana Milne Edwards and Haime, 1849. The latter species shows great morphological variation in paliform lobes, corallite size, and colony form and has been found living off Rio de Janeiro [Laborel, 1971] and in the Pliocene of Venezuela [Weisbord, 1968] and Florida [Weisbord, 1974]. In view of the variation and distribution of P. americana and the unique but unfortunately vague distributional record of P. fuegoensis it is possible that they will be synonymized when more specimens become available.

Material. Holotype

Types. The holotypic colony is deposited at the Museum of Comparative Zoology (5390). Type-locality: Tierra del Fuego (? fossil).

Distribution. Known only from type-locality.

Family OCULINIDAE Gray, 1847
Genus Bathelia Moseley, 1881

Diagnosis. Colonial, extratentacular budding forming dendroid coralla. Coenosteum dense. Septal edges smooth. Crown of pali before S_3; columella of irregular ribbons. Type-species: Bathelia candida Moseley, 1881, by monotypy.

6. Bathelia candida Moseley, 1881
Plate 3, figs. 1-3

Bathelia candida Moseley, 1881, pp. 177, 178, pl. 8, figs. 1-6.--Wells, 1958, p. 262.--Cairns, 1979, p. 206.

Description. Colony dendroid, corallites arranged in opposite and alternating fashion on branch. Extratentacular budding most common; however, intratentacular budding also occurring. May have three branches originating at one calice. Maximum size of colony unknown. Branches robust (about 1 cm in diameter) and solid; no dissepiments present. Coenosteum smooth and very finely granulated. Thin, shallow coenosteal striae bordering wide, flat costae corresponding to all septa. Calices round to slightly elliptical, 6-10 mm in diameter, projecting obliquely several millimeters from branch.

Septa hexamerally arranged in four cycles. S_1 and S_2 equal in size and extending to columella; S_3 and S_4 progressively smaller. All septa narrow and slightly exsert, extending over thickened calicular edge as low ridges. Inner septal edges straight; those of S_1 and S_2 usually entire, those of S_3 and S_4 usually dentate. Septal granulation variable, ranging from abundant, small, fine granules to sparse larger, blunt granules. Lower inner edges of S_3 bearing tall, narrow pali, these bordering columella and terminating at same height as columellar papillae. Pali usually distinguished from papillae by their larger size and elliptical to rectangular shape in cross section. Columella composed of 5-15 tall, slender papillae, these irregularly round in cross section. Pali and columellar papillae granulated. Because of narrow septa, fossa rather wide but not very deep, being filled in with pali and columella.

Discussion. Bathelia is a monotypic genus and has been reported from only one locality previously.

Material. Vema sta. 17-14, USNM 47512. Calypso sta. 171, USNM 47513 and SME. Following WH records (H. Zibrowius, personal communication, 1979): sta. 215/66, sta. 142/71, sta. 191/71, sta. 197/71, sta. 328/71, sta. 329/71, sta. 331/71 (all WH specimens deposited at ZIZM). Syntypes.

Types. The syntype branches, from Challenger sta. 320, are deposited at the British Museum. Type-locality: 37°17'S, 53°52'W (off Rio de la Plata); 1097 m.

Distribution. Off southern South America from Rio Grande, Brazil, to Cabo Tres Puntas, Argentina; off Peninsula de Taitao, Chile (Map 1). Depth range: 500-1250 m.

Genus Madrepora Linnaeus, 1758

Diagnosis. Colonial, extratentacular budding forming dendroid colonies. Coenosteum dense, no costae, corallites filled internally by stereome. No pali; columella spongy or absent. Type-species: Madrepora oculata Linnaeus, 1758, by subsequent designation [Verrill, 1901].

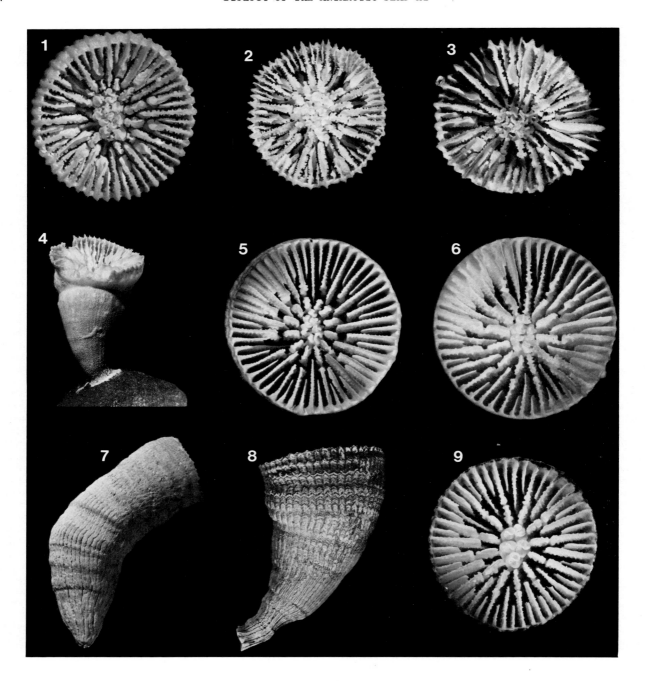

Plate 4. Caryophyllia

1-4. Caryophyllia antarctica Marenzeller: 1, USNM 47304, Eltanin sta. 1933, GCD
      = 14.6 mm; 2, USNM 53414, Yelcho 2-11, GCD = 13.1 mm; 3, specimen identified
      as Caryophyllia clavus by Thomson and Rennet [1931], Australian Museum G
      13536, sta. 10, GCD = 17.9 mm; 4, USNM 47302, Glacier sta. 1, H = 16.0 mm.

5-9. Caryophyllia squiresi n. sp.: 5, paratype, USNM 47161, Eltanin sta. 558,
      GCD = 15.6 mm; 6, paratype from same lot, GCD = 17.1 mm; 7, paratype from
      same lot, H = 38.3 mm; 8, 9, holotype, USNM 47160, Eltanin sta. 558, GCD =
      15.6 mm, H = 31.0 mm, theca coated with ammonium chloride.

7. Madrepora oculata Linnaeus, 1758
Plate 3, figs. 4-6

Madrepora oculata Linnaeus, 1758, p. 798.--von Marenzeller, 1904b, p. 79.--Eguchi, 1968, p. C-29, pl. C-8, figs. 1-9.--Zibrowius, 1974a, pp. 762-766, pl. 2, figs. 2-5; 1980, pp. 36-40, pl. 13, figs. A-P.--Cairns, 1979, pp. 39-42, pl. 3, fig. 2, pl. 4, fig. 5, pl. 5, figs. 1-3.
Amphihelia oculata; Milne Edwards and Haime, 1857, p. 119.--von Marenzeller, 1904a, p. 308, pl. 14, figs. 1, 1b.
Amphihelia ramea; Duncan, 1873, p. 326, pl. 44, figs. 1-3, pl. 45, figs. 4-6, pl. 46, figs. 1-19.
Lophohelia candida Moseley, 1881, pp. 179, 180, pl. 9, figs. 6-13.
? Madrepora vitiae Squires and Keyes, 1967, p. 22, pl. 1, figs. 4-8.

Description. Colony bushy or flabellate, formed by extratentacular budding. End branches having sympodial arrangement of corallites, measuring between 2.3 and 4.0 mm in diameter; diameter of attached base up to 2 cm. Calices round, 2.4-3.8 mm in diameter, exsert on end branches, recessed or flush with coenosteum toward base. Coenosteum smooth, extremely finely granulated; costae and coenosteal striae rare.

Septa hexamerally arranged in three cycles. S_1 equal to or larger than S_2; S_3 much smaller, sometimes rudimentary. Inner edges of septa straight, sometimes thickened near columella. Septal faces covered by granules, sometimes twice as high as septal thickness. Fossa variable in depth, usually dependent on age of corallite, older corallites having shallower fossae. Columella variable, usually papillose, sometimes absent.

Discussion. More complete synonymies and descriptions are given by Zibrowius [1974a, 1980] and Cairns [1979]. Zibrowius [1974a] lists the nominal species of Madrepora and discusses their relationship to M. oculata.

Madrepora oculata is a widespread and extremely variable species. Characters that are subject to variation, sometimes within the same colony, include frequency of branching, intercorallite distance, coenosteum texture and color, relative septal sizes, septal granulation, fossa depth, and development of columella.

A closely related, if not identical, species, M. vitiae Squires and Keyes, 1967, was also collected off New Zealand (Eltanin stations 1814, 1816, and 1818; NZOI station C-642) (Map 2). The only difference between the two is that M. vitiae usually has paliform lobes, sometimes quite thick, before the S_2. However, a branch of a topotypic specimen of M. vitiae has corallites with and without pali, and some calices have a variable number of paliform lobes (1-6). The T-shaped inner septal edges mentioned by Squires and Keyes [1967] were not observed in specimens collected from the type-locality (NZOI station B-314) or in specimens from five other lots near the type-locality. They may have been referring to the slight thickening of the inner septal edges, which is common in M. oculata.

Zibrowius's [1974a] M. oculata from off Île Saint-Paul, Indian Ocean, is similar to M. vitiae; however, he did not consider the presence of P_2 as a specific difference. If the presence or absence of P_2 is considered to be of no specific

value, then M. vitiae may be dropped to a form.

Material. Eltanin sta. 254, USNM 47500; sta. 1346, USNM 47499; sta. 1403, USNM 47501; sta. 1416, USNM 47665; sta. 1422, USNM 47497; sta. 1814, USNM 47502; sta. 1816, USNM 47498; sta. 1818, USNM 47504. NZOI sta. C-642, USNM 47514; sta. D-6, USNM 47503. Specimens listed by Cairns [1979], USNM; topotypic specimens of M. vitiae from NZOI sta. B-314, type lot, USNM 47515. Syntypes of L. candida.

Types. The types of M. oculata are lost. Type-locality: off Sicily and Tyrrhenian Sea, Mediterranean. Syntypes of L. candida are deposited at the British Museum (1880.11.25.95). Type-locality: off Sombrero island, Lesser Antilles; 823 m. The holotype of M. vitiae is deposited at the New Zealand Oceanographic Institute (17). Type-locality: off Cape Farewell, New Zealand; 230-251 m.

Distribution. According to Zibrowius [1974a, p. 776], distribution of M. oculata worldwide outside of polar seas. Three of above-mentioned records extend the southernmost distribution of M. oculata to Subantarctic waters: Hjort Seamount, a seamount in the Subantarctic South Pacific, and a seamount in the Drake Passage (Map 2). Worldwide depth range: 80-1500 m; Subantarctic records: 549-833 m.

Suborder CARYOPHYLLIINA Vaughan and Wells, 1943
Superfamily CARYOPHYLLIICAE Gray, 1847
Family CARYOPHYLLIIDAE Gray, 1847
Subfamily CARYOPHYLLIINAE Gray, 1847
Genus Caryophyllia Lamarck, 1801

Diagnosis. Solitary; ceratoid, turbinate, or subcylindrical; fixed or free. Septotheca usually costate. Pali opposite S_3 in one crown (or before second group of septa when hexameral symmetry obscured). Columella fascicular, formed of twisted ribbons. Type-species: Madrepora cyathus Ellis and Solander, 1786, by subsequent designation [Broderip, 1828].

8. Caryophyllia antarctica Marenzeller, 1904
Plate 3, figs. 7-9; Plate 4, figs. 1-4

Caryophyllia antarctica Marenzeller, 1903, p. 1 (nomen nudum); 1904a, pp. 293, 294, pl. 16, figs. 7, 7d.--Pax, 1910, pp. 65, 66, pl. 11, fig. 1.--Gravier, 1914b, pp. 129, 130, pl. 1, figs. 7, 8.--Wells, 1958, pp. 267, 268, pl. 2, figs. 3, 4.--Squires, 1961, p. 20; 1962b, pp. 13, 14, 16, 17, pl. 1, figs. 11, 12; 1969, pp. 16, 17, pl. 6, map 1.--Eguchi, 1965, pp. 7, 8, pl. 1a, 1b. --Cairns, 1979, p. 206.
Caryophyllia clavus; Thomson and Rennet, 1931, p. 40.
Caryophyllia arcuata; Gardiner, 1939, pp. 331, 332.

Description. Corallum ceratoid to trochoid, usually straight, attached. Pedicel diameter one fifth to one third of GCD, expanding only slightly at substrate. Largest corallum examined 28.6 x 26.2 mm in CD and 36.5 mm tall; however, more typical coralla 10-15 mm in GCD and 15-20 mm tall. Theca usually smooth, porcelaneous, sometimes with flat, equal costae bordered by shallow intercostal striae. Costal granules rare; when present, low and rounded. Calice round to elliptical.

Septa usually hexamerally arranged in four cycles. S_1 and S_2 equal in size and exsertness; S_3 and S_4 progressively smaller. Larger coralla with up to 90 septa, accommodated by increase in number of half systems and acceleration of higher-cycle septa instead of by addition of another cycle of smaller septa. Inner edges of S_1, S_2, and S_4 slightly sinuous, those of S_3 and P_3 very sinuous. Septal granulation prominent, usually arranged in widely spaced rows on septal undulations oriented parallel to septal edge. Individual granules sometimes quite tall, with rounded, clavate, bifid, or squared-off tops. Granules usually fused into low, distinct carinae, these having continuous or serrated (beaded) upper edges. Carinae especially well developed near inner septal edges.

Fossa shallow. Pali of varying widths (up to width of S_3) stand before S_3; each separated from its corresponding septum by deep, narrow notch. Pali sometimes split into two smaller lobes. Pali sometimes present before S_1 and S_2. Palar granulation similar to that of septa but more prominent; carinae running obliquely across palus. Columella composed of 4-20 discrete, slender, twisted ribbons aligned in greater axis of calicular ellipse.

Discussion. C. antarctica is distinguished by its distinctive septal ornamentation of carinae and squared-off granules. Thomson and Rennet's [1931] C. clavus is a typical specimen of C. antarctica.

Material. Eltanin sta. 138 (4), USNM 45677; sta. 416 (1), USNM 47307; sta. 428 (6), USNM 45670; sta. 678 (1), USNM 47291; sta. 992 (1), USNM 47306; sta. 1067 (1), USNM 47296; sta. 1081 (2), USNM 47301; sta. 1082 (4), USNM 47297; sta. 1084 (8), USNM 47317; sta. 1870 (11), USNM 47309; sta. 1883 (2), USNM 47289; sta. 1922 (1), USNM 47294; sta. 1930 (1), USNM 47311; sta. 1931 (1), USNM 47285; sta. 1933 (11), USNM 47304; sta. 1995 (25), USNM 47316; sta. 1966 (45), USNM 47315; sta. 2007 (1), USNM 47300; sta. 2022 (3), USNM 47205; sta. 2079 (3), USNM 47293; sta. 2104 (1), USNM 47668; sta. 2106 (1), USNM 47286; sta. 2119 (1), USNM 47288; sta. 5765 (6), USNM 47284. Islas Orcadas sta. 876-118 (1), USNM 47303. Hero sta. 721-849 (2), USNM 47308; sta. 731-1812 (1), USNM 47318. Yelcho sta. 2-11 (5), USNM 53414. Glacier sta. 1 (1), USNM 47302. Edisto sta. 21 (1), USNM 47920; sta. 31 (2), USNM 47293; sta. 36 (1), USNM 47310. Staten Island sta. 21 (1), USNM 47298. Atka sta. 23 (42), USNM 47313. Burton Island sta. 3 (20), USNM 47305. EW sta. 4 (1), USNM 47287; sta. 35 (5), USNM 47314. GLD sta. 15 (1), USNM 47299. Specimens (3) identified as Caryophyllia clavus by Thomson and Rennet [1931], Australian Museum G 13536; specimens of Wells [1958] from the following Discovery stations: sta. 39 (12), H 43; sta. 41 (3), H 46; sta. 40 (1), H 47 (all deposited at the South Australian Museum, Adelaide). Syntypes.

Types. Four syntypes collected by the Belgica (station 290 (3) and station 569 (1)) are deposited at the Brussels Museum. Another two syntypes from Valdivia station 127 are deposited at the Museum für Naturkunde, Berlin (5067). Type-locality: near Peter I Island, Antarctica, and off Bouvetøya; 567 m.

Distribution. Endemic to the Antarctic region, probably circumpolar. Squires's [1969] records from Subantarctic South America undocumented (Map 3). Depth range: 87-1435 m.

9. Caryophyllia squiresi n. sp.
Plate 4, figs. 5-9

Caryophyllia sp. A Squires, 1969, p. 17 (part: 3 of 4 South American records only), pl. 5, map 1.--Cairns, 1979, p. 206.

Description. Corallum ceratoid, sometimes becoming cylindrical, often bent near base but rarely by more than 40°. Attached by narrow pedicel usually 2.9-3.4 mm in diameter (18-25% GCD of adult corallum). Holotype 15.6 x 14.5 mm in CD, 3.0 mm in PD, and 31.0 mm tall. Largest specimen 18.6 mm in GCD. Costae equal, smooth, porcelaneous, and bordered by very thin, intercostal striae. Calice slightly elliptical.

Septa hexamerally arranged in four cycles. S_1 and S_2 equal in size; S_3 and S_4 progressively smaller. Pairs of smaller S_5 present in larger coralla, flanking S_4; one specimen has 10 S_5, or 58 septa. These S_4 then enlarged to almost size of an S_3. Septa not exsert. Inner septal edges straight, except for those of S_3, these sometimes sinuous. Septal granulation sparse and usually nonlinear. Granules usually low and blunt, rarely squared off, and never fused into carinae.

Fossa moderately deep. Tall, narrow pali present before S_3, each separated from its corresponding septum by deep, narrow notch. Palar margins usually sinuous and granulation usually more prominent than that of septa. Pali (usually 12) often forming distinct crown but may merge indistinguishably with columella. Columella composed of 3-10 discrete, twisted ribbons aligned in plane of greater axis of calicular ellipse,

Discussion. This species corresponds to at least one lot of specimens identified by Squires (Eltanin station 558) and referred to by him [Squires, 1969] as Caryophyllia sp. A. Two more of his records are consistent with specimens at the United States National Museum; however, the other 17 records [Squires, 1969, pl. 6, map 1] from off western South America, Antarctica, and New Zealand are undocumented. Until these specimens are found and verified, the distribution of C. squiresi will remain as follows: off Tierra del Fuego and off the Falkland Islands (Islas Malvinas).

C. squiresi is similar to C. antarctica but can be distinguished by its lack of septal carinae, deeper fossa, less sinuous inner septal edges, and less exsert septa.

Etymology. This species is named in honor of D. F. Squires, who has done much to advance the knowledge of Antarctic Scleractinia and who first recognized this species.

Material. Eltanin sta. 339 (4), USNM 47516. Vema sta. 15-PD9 (1), USNM 47517, and (1), AMNH. Types.

Types. The holotype, collected at Eltanin station 558, is deposited at the United States National Museum (47160). Twenty-two paratypes, collected at Eltanin station 558, are deposited at the United States National Museum (47161), and another specimen from the same station is deposited at the British Museum (1979.11.1.1). Type-locality: 51°58'S, 56°38'W (off East Falkland island); 646-845 m.

Distribution. See discussion (Map 3). Depth range: 406-659 m.

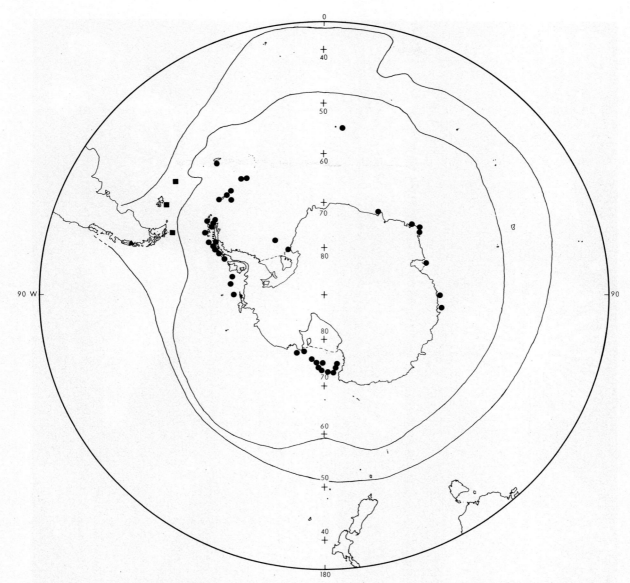

Map 3. Distribution of <u>Caryophyllia</u> <u>antarctica</u> (solid circles), <u>Caryophyllia</u> <u>squiresi</u> (solid squares), and <u>Caryophyllia</u> <u>clavus</u> var. <u>smithi</u> (solid triangle).

10. <u>Caryophyllia profunda</u> Moseley, 1881
Plate 5, figs. 1-5

<u>Caryophyllia</u> <u>profunda</u> Moseley, 1881, pp. 138, 139
(part: specimen from Cape Verde Islands is <u>C.</u>
<u>cyathus</u>), pl. 1, figs. 6, 6b.--Not <u>C.</u> <u>profunda</u>;
Jourdan, 1895, pp. 10, 11 (is <u>C.</u> <u>cyathus</u> Ellis
and Solander, 1786).--von Marenzeller, 1904a, p.
298.--Gardiner, 1913, pp. 688, 689.--Not <u>C.</u> <u>pro-
funda</u>; Gravier, 1920, p. 28 (is <u>Caryophyllia</u>
<u>foresti</u> Zibrowius, 1980).--Gardiner, 1929a, p.
126; 1939, p. 331.--Ralph, 1948, p. 108, fig. 2
(top).--Squires, 1958, p. 44; 1960, pp. 196,
198-200, pl. 34, figs. 5-7, pl. 35, figs. 9-11;
1962b, pp. 13-15, pl. 1, figs. 13, 14; 1964a, p.
11; 1969, pp. 16, 17, pl. 6, map 1.--Ralph and
Squires, 1962, pp. 3, 6, 7, pl. 1, figs. 8-11.--
Squires and Keyes, 1967, pp. 15, 17, 23, pl. 2,
figs. 1-4.--Zibrowius, 1974a, pp. 751-755, pl.
1, figs. 1-10.--Beurois, 1975, p. 46, photo
13.--Cairns, 1979, p. 206.

<u>Caryophyllia</u> <u>cyathus</u>; von Marenzeller, 1904a, p.
295, pl. 16, figs. 6, 6a.--Hoffmeister, 1933, p.
14, pl. 4, figs. 4, 5.--Gardiner, 1939, pp. 330,
331.--Squires, 1961, p. 17.

<u>Caryophyllia</u> <u>planilamellata</u> Dennant, 1906, pp.
157, 158, pl. 6, figs. 4a, 4b.--Squires, 1961,
p. 18.

<u>Caryophyllia</u> <u>clavus</u>; Wells, 1958, p. 265, pl. 1,
figs. 12, 13.

<u>Caryophyllia</u> cf. <u>C.</u> <u>maculata</u>; Ralph, 1948, p. 108,
fig. 2 (bottom, right).--Ralph and Squires, 1962,
pp. 3, 7, pl. 2, figs. 1, 2.--Squires and Keyes,
1967, pp. 15, 17, 23, pl. 2, figs. 4, 5.

<u>Description</u>. Corallum trochoid to cylindrical,
straight to slightly bent; strongly attached by
broad, encrusting base. Pedicel variable in diameter, ranging from 20 to 70% of GCD. Pedicel
usually greatly increased in diameter by concentric
layers of external stereome. Large specimens up
to 41 mm in GCD and 50 mm tall. Individual coralla

Plate 5. Caryophyllia

1-5.   Caryophyllia profunda Moseley: 1, USNM 47519, Eltanin sta. 1718, H = 48.7
       mm; 2, specimen from same lot, GCD = 28.2 mm; 3, specimen reported by
       Gardiner [1939], BM 1939.7.20.207, Discovery sta. 1187, GCD = 25.1 mm; 4,
       specimen identified as C. cyathus by Gardiner [1939], BM 1939.7.20.210,
       Discovery sta. 190, GCD = 16.2 mm; 5, syntype, BM 1880.11.25.36, Challenger
       sta. 135, GCD = 25.5 mm.
6-9.   Caryophyllia eltaninae n. sp.: 6, 7, holotype, USNM 47162, Eltanin sta.
       671, H = 35.6 mm, GCD = 26.8 mm; 8, USNM 47163, Eltanin sta. 671, GCD =
       25.8 mm; 9, specimen from same lot, H = 28.0 mm.

sometimes clumping into pseudocolonial arrangement. Calice round in young specimens, becoming elliptical in larger specimens. Theca porcelaneous, finely granulated, and often brownish. Costae usually flat and equal, although C_{1-3} sometimes moderately ridged. Septa hexamerally arranged in five cycles. S_1 and S_2 equal in size and highly exsert; higher-cycle septa progressively smaller and less exsert. Calices with more than 96 septa rare, but those with less than 96 septa and 24 pali common, roughly a function of smaller calicular diameter. Inner edges of all septa straight, except those of S_4, these sometimes slightly sinuous. Septal granulation variable, usually consisting of extremely fine, low granules, but sometimes larger, blunt granules; never arranged in carinae.

Fossa moderately deep. Narrow pali occurring before S_4; each separated from its corresponding septum by a deep and narrow to shallow and broad notch. Pali sometimes bilobed or trilobed. Columella variable, composed of several linearly arranged, twisted ribbons; or a fused mass of twisted ribbons generally aligned in greater calicular axis; or a labyrinthiform arrangement of modified twisted ribbons.

Remarks. Only Moseley [1881] recorded observations of the living coral. He stated that the ground color of the polyp was transparent blue, encircled by a sulphur-yellow margin at the calicular edge. The stomadeum was white or vermillion, and the short tentacles were red knobbed.

In one of the few papers that document growth rates for deepwater corals, Squires [1960] estimated the growth rate for this species as 0.88-2.02 mm/year in height. He also hypothesized on features characteristic of cessation of growth (or maximum size), such as lobation of pali and septa; increased thickness of septa and pedicel; and coarsened septal, palar, and costal ornamentation.

Discussion. Gardiner's [1939] record of C. profunda from Discovery station 190 is the only continental Antarctic record for the species and produces an unusual distribution pattern, which includes predominantly cold temperate records, two marginal Subantarctic records (Tristan and Gough islands), and Gardiner's single Antarctic record. Zibrowius [1974a, p. 754] distinguished Gardiner's Antarctic specimen from typical C. profunda by its narrower and deeper notches between the septa and pali and the more vertical edges of its pali. Among the specimens that I have examined, I find these characters to be within the range of variation for the species and, in general, not of specific value. Assuming that no labeling errors were made, the somewhat anomalous distribution of C. profunda must stand.

C. profunda is easily distinguished from other Antarctic Caryophyllia by its greater size and the presence of five cycles of septa with pali before the fourth cycle.

Material. Eltanin sta. 1403 (1), USNM 47518; sta. 1718 (37), USNM 47519; sta. 1814 (1), USNM 47520. Specimens (8) identified as C. profunda and C. cyathus by Gardiner [1939], BM 1939.7.20.202-203, 207-213; some specimens of Squires and Keyes [1967], i.e., B-489 (5), C-690 (4), C-703 (5), all at USNM; some species of Zibrowius [1974a], i.e., AMS-66 (1), AMS-1474, off Île Amsterdam, 80 m, Jan. 1972, all at USNM. Eleven syntypes.

Types. Approximately 20 syntypes of C. profunda, collected at Challenger station 135, are deposited at the British Museum (1880.11.25.36, 1880.11.25.241, 1889.7.8.1-5). The syntype fragment from Cape Verde Islands is C. cyathus [see Zibrowius, 1974a]. Type-locality: 37°01'50"S, 12°19'10"W (off Nightingale Island, Tristan da Cunha Group); 183-274 m. At least one syntype of C. planilamellata Dennant is deposited at the Australian Museum (G 12057).

Distribution. Circumpolar in southern temperate waters: off South Africa, Ile Saint-Paul and Île Amsterdam, South Australia, New Zealand, and Chatham Island; Subantarctic islands of Tristan and Gough; off Hugo Island, Palmer Archipelago. Squires's [1969] Subantarctic records from off South America and the Macquarie Ridge are unsubstantiated (Map 4). Most common between 80 and 250 m; confirmed range: 35-1116 m.

11. Caryophyllia eltaninae n. sp.
Plate 5, figs. 6-9

Gardineria lilliei; Gardiner, 1939, pp. 328, 329 (part: two specimens from Discovery sta. 160).

Description. Corallum ceratoid to trochoid, attached or free. If attached, corallum usually straight, with reinforced pedicel up to 38% of GCD; if free, corallum often slightly bent, with an eroded base as small as 10% of GCD. Holotype (largest specimen) 26.8 x 23.5 mm in CD, 6.1 mm in PD, and 35.6 mm tall. Costae usually nongranulated and porcelaneous, bordered by thin intercostal striae. C_{1-3} slightly ridged in some specimens. Calice round to elliptical.

Septa hexamerally arranged in five cycles. S_1 and S_2 equal in size and exsertness; septa of higher cycles progressively smaller. Full fifth cycle attained at about 13.5 mm CD. S_1 and S_2 extending to columella; S_5 rudimentary, with irregularly dentate inner edges. Inner edges of all septa and pali straight. Septal granules low and blunt, never arranged in carinae.

Pali (12) occurring before S_3 and variable in shape; usually tall and narrow, but sometimes triangular or twisted like a columellar ribbon. In about one fourth of specimens examined, paliform lobes also present on S_4, often in form of small, horizontally projecting lobe directed at, and sometimes merging with, adjacent P_3. Palar granulation more prominent than that of septa. Columella variable, composed of 4-30 discrete, twisted ribbons or fused mass of twisted elements aligned in greater axis of calicular ellipse.

Discussion. Gardiner's [1939] misidentification of this species as G. lilliei is a result of the small size of his species (CD = 9.8 x 9.8, 9.0 x 10.0 mm), which at this stage could be confused with Gardineria. His specimens were just beginning to form pali and S_5; both specimens had only one palus, corresponding to the half systems where S_5 had formed.

C. eltaninae is unusual in that its pali occur before the antipenultimate septal cycle, not the penultimate as is common in most Caryophyllia. This is a character shared with the Caribbean C. paucipalata Moseley, 1881. It is further distinguished from C. antarctica and C. squiresi by its straight inner septal edges.

Etymology. This species is named after the R/V

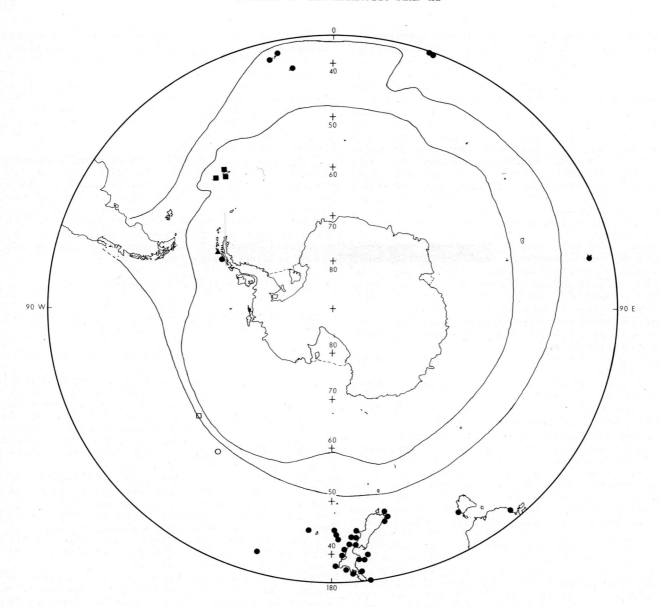

Map 4. Distribution of _Caryophyllia profunda_ (solid circles), _Caryophyllia eltaninae_ (solid squares), _Caryophyllia mabahithi_ (solid triangle), _Cyathoceras_ A (open circle), and _Cyathoceras irregularis_ (open square).

Eltanin, from which many of the specimens used in this study were collected, including the holotype of this species.

Material. _Eltanin_ sta. 678 (2), USNM 47486; sta. 1535 (1), USNM 47485. _Islas Orcadas_ sta. 575-8 (1), USNM 47487; sta. 575-10 (3), USNM 47490; sta. 575-12 (3), USNM 47489; sta. 575-14 (2), USNM 47488; sta. 575-17 (5), USNM 47484; sta. 575-93 (2), USNM 47491. Specimens (2) identified as _G. lilliei_ by Gardiner [1939], BM 1939.7.20.286-287. Types.

Types. The holotype, collected at _Eltanin_ station 671, is deposited at the United States National Museum (47162). Thirty-nine paratypes from _Eltanin_ station 671 (number 47163) and 11 paratypes from _Islas Orcadas_ station 575-11 (number 47164) are also deposited at the United States National Museum. One paratype from _Eltanin_ station 671 is also deposited at the British Muse-

um (1979.71.2.1.). Type-locality: 54°41'S, 38°38'W (off southwest South Georgia); 220-320 m.

Distribution. Known only from the shelf and slope off the western half of South Georgia and off Shag Rocks (Map 4). Depth range: 101-261 m, except for one record at 778-814 m.

12. _Caryophyllia mabahithi_ Gardiner and Waugh, 1938
Plate 6, figs. 1-5

Caryophyllia mabahithi Gardiner and Waugh, 1938, pp. 178, 179, text fig. 1, pl. 3, fig. 6.--Gardiner, 1939, p. 332.--Wells, 1958, p. 262.--Squires, 1961, p. 21; 1962b, pp. 14, 16, 17, pl. 1, figs. 15, 16.

Description. Corallum free, trochoid, generally curved between 45° and 90°. Base 1.2-1.5 mm in

Plate 6. Caryophyllia and Cyathoceras

1-5. Caryophyllia mabahithi Gardiner: 1-3, specimens reported by Gardiner [1939], BM 1939.7.20.247-248, Discovery sta. 182, GCD = 9.0 mm, 8.6 mm; 4, 5, syntype, BM 1950.1.9.561-586, John Murray sta. 34, GCD = 9.8 mm, H = 9.3 mm.

6-9. Cyathoceras irregularis n. sp.: 6, 7, paratype, USNM 47166, Eltanin sta. 1346, H = 18.5 mm, GCD = 15.8 mm; 8, 9, holotype, USNM 47165, Eltanin sta. 1346, H = 14.9 mm, GCD = 13.7 mm.

diameter, usually showing hexameral partitioning of basal plate caused by first-cycle septa. An average-sized specimen is 9.3 x 8.3 mm in CD and 9.6 mm tall; however, Gardiner reported specimens 12 x 10 mm in CD and 13 mm tall. Costae flat or slightly ridged, covered by very low, rounded granules arranged three or four across width of a costa near calice.

Septa octamerally arranged in three cycles; additional septa rare. S_1 highly exsert and extending almost to columella; S_2 and S_3 progressively smaller and much less exsert. S_1 and S_3 with slightly sinuous inner edges, those of S_2, however, more sinuous. Crown of eight pali occuring before S_2, each palus separated from its corresponding septum by a deep, narrow notch. Palar margins very sinuous. Septal granulation variable, consisting of pointed granules ranging from 0.5 to 2.0 times septal width in height; palar granulation usually even more prominent. Columella composed of one to seven broad, twisted ribbons fused among themselves and to inner edges of pali.

Discussion. The three specimens reported by Gardiner [1939] from off the Palmer Archipelago are, by direct comparison, indistinguishable from syntypes of C. mabahithi from the Gulf of Aden. This disjunct distribution, both geographic and bathymetric, is very unusual and inexplicable.

Two other species of octameral Caryophyllia are known: C. octopali Vaughan, 1907 (Hawaii), and C. barbadensis Cairns, 1979 (Barbados). They are both readily distinguished by their subcylindrical, firmly attached coralla and smaller calices.

Material. Specimens (3) of Gardiner [1939], BM 1939.7.20.246-248. Syntypes from John Murray sta. 34 (26 specimens).

Types. Sixty-four syntypes of C. mabahithi are deposited at the British Museum. Those from station 34 are numbered 1950.1.9.561-586. Two syntypes from this lot have been permanently transferred to the United States National Museum (48299). Type-locality: Gulf of Aden and Chagos Archipelago; 655-1022 m.

Distribution. Gulf of Aden; Chagos Archipelago; off Anvers Island, Palmer Archipelago (Map 4). Depth range: 278-1022 m.

Genus Cyathoceras Moseley, 1881

Diagnosis. Solitary, ceratoid to turbinate, fixed. Septotheca usually costate. No pali. Columella fascicular, composed of several twisted ribbons. Type-species: Cyathoceras cornu Moseley, 1881, by subsequent designation [Faustino, 1927].

13. Cyathoceras irregularis n. sp.
Plate 6, figs. 6-9

Description. Corallum ceratoid, straight to irregularly bent, attached by thin, encrusting, slightly expanded base. Pedicel one fourth to one third of GCD. Holotype 13.7 x 10.5 mm in CD and 14.9 mm tall; largest specimen 15.4 x 14.6 mm in CD and 18.4 mm tall. Theca smooth and procelaneous, covered by very small, low granules. Costae occurring near calicular edge and base, if at all. Calice round to elliptical.

Septa hexamerally arranged in four or five cycles; however, fifth cycle never complete: largest specimen with 72 septa. Up to 48 septa stage,

S_1 and S_2 equal in size, moderately exsert, and extending to columella. S_3 and S_4 progressively narrower and less exsert. With addition of S_5, septal arrangement becomes irregular. May have one or two pairs of S_5 in each system. If only one pair present, enclosed (flanked) S_4 invariably larger than unflanked S_4. If both pairs of S_5 present, both S_4 enlarged, reaching almost as far toward columella as S_3. Occurrence of all degrees of S_5 development in one corallum possible, i.e., systems with 0, 1, or 2 pairs of S_5, making interpretation of septal cycles confusing. Inner edges of S_{1-3} broadly sinuous, corresponding to transverse septal undulations, but inner edges of higher-cycle septa straight. Large, blunt granules present on septal faces, usually arranged in lines along crests of septal undulations.

Columella large, composed of numerous slender, twisted ribbons usually fused into solid mass.

Remarks. All specimens examined were attached to dead coral, usually Solenosmilia variabilis. The type-locality is a seamount or ridge, which supports a presumed deepwater coral bank composed primarily of S. variabilis. C. irregularis is similar to Cyathoceras squiresi Cairns, 1979, in its association with the framework coral of deepwater banks. C. squiresi is often attached to Enallopsammia profunda, a common constituent of western Atlantic deepwater banks.

Discussion. For the purposes of this comparison the following ten species are considered valid Cyathoceras: C. cornu Moseley, 1881; C. rubescens Moseley, 1881; C. tydemani Alcock, 1902; C. diomedeae Vaughan, 1907; C. niinoi Yabe and Eguchi, 1942; C. foxi Durham and Barnard, 1952; C. woodsi Wells, 1964; C. squiresi Cairns, 1979; C. avis (Durham and Barnard, 1952); and C. hoodensis (Durham and Barnard, 1952). C. quaylei Durham, 1947, is herein transferred to Labyrinthocyathus, on the basis of an examination of the paratypes (USNM 547417), which have columellas composed of interconnected lamellae. C. irregularis can be distinguished from all of these species by its fused columella, composed of closely united, poorly defined, twisted ribbons; other Cyathoceras have well-defined columellar elements. It is also distinguished by the irregularity of development of the fifth-cycle septa.

Etymology. The specific name irregularis refers to the irregular manner in which the S_5 are added.

Material. Types.

Types. The holotype, collected at Eltanin station 1346, is deposited at the United States National Museum (47165). Seven paratypes, also collected at Eltanin station 1346, are deposited at the United States National Museum (47166), and one paratype from this station is deposited at the British Museum (1979.11.3.1). Type-locality: 54°49'S, 129°48'W (seamount or ridge on Heezen fracture zone of Eltanin fracture zone system); 549 m.

Distribution. Known only from type-locality (Map 4).

14. Cyathoceras sp. A
Plate 7, figs. 1, 2

Description. Corallum ceratoid, straight, firmly attached. This specimen 8.9 x 8.6 mm in CD, 5.5 mm

Plate 7. Cyathoceras, Stephanocyathus, and Aulocyathus

1, 2.   Cyathoceras sp. A: USNM 47521, Eltanin sta. 17-6, GCD = 8.9 mm, H = 16.5 mm.

3-6.   Stephanocyathus platypus (Moseley): 3, 4, syntype, BM 1880.11.25.57, Challenger sta. 164, CD = 48.4 mm; 5, 6, USNM 47522, Eltanin sta. 1718, CD = 53.0 mm.

7-9.   Aulocyathus recidivus (Dennant): 7, USNM 47524, NZOI sta. C-734, CD = 8.8 mm; 8, 9, specimen from same lot, CD = 10.3 mm, H = 17.8 mm, theca coated with ammonium chloride.

in PD, and 16.5 mm tall. Theca thin, porcelaneous, not granulated or costate. Thin, dull white longitudinal striae, corresponding to interseptal spaces, faintly distinguishable on parts of theca.

Septa decamerally arranged in three septal sizes. Ten primaries are the largest septa, highly exsert (1.5 mm above calicular edge), and extending to columella. Ten secondaries are slightly exsert and extend two thirds of distance to columella. Twenty tertiaries barely exsert and extending one fourth of distance to columella. Inner edges of primaries and secondaries broadly sinuous (amplitude of undulations high and period long); tertiaries less corrugated. Long, prominent carinae on septal faces of primary and secondary septa, corresponding to crests of the broad septal corrugations. Tertiaries bearing only sparse septal granules. Columella composed of three broad, nongranulated, twisted ribbons, typical in shape for Cyathoceras. Carinae, septa, and columella all translucent.

Discussion. The description above is based on only one specimen, which probably represents a new species; however, because of the lack of additional specimens to provide some indication of variation, it is not named here. This specimen was collected just north of the Subantarctic region as defined by Hedgpeth [1969] and is included here only because of its proximity to the Subantarctic region.

Four of the eleven previously listed Cyathoceras have decameral symmetry: C. avis (Durham and Barnard, 1952); C. hoodensis (Durham and Barnard, 1952); C. woodsi Wells, 1964; and C. squiresi Cairns, 1979. Cyathoceras A is distinguished from the first two eastern Pacific species by its firm attachment to the substrate and its lack of costae. It is easily distinguished from C. woodsi by its larger size and septal carinae; however, it is similar to C. squiresi, especially in size, septal granulation, and septal sinuosity. The main points of difference are that Cyathoceras A has a thinner theca, exsert septa, and no costae. Furthermore, it is found far deeper than any other species of Cyathoceras.

Material. Eltanin sta. 17-6 (1), USNM 47521.

Distribution. Known from 52°10'S, 142°10'W (Tharp fracture zone of Eltanin fracture zone system) (Map 4). Depth range: 2305-2329 m.

Genus Stephanocyathus Seguenza, 1864

Diagnosis. Solitary, patellate, free. Costae usually present. Paliform lobes usually present on all septa. Columella trabecular, papillose, or fused on surface. Type-species: Stephanocyathus elegans Seguenza, 1864, by subsequent designation [Wells, 1936].

15. Stephanocyathus platypus (Moseley, 1876)
Plate 7, figs. 3-6

Ceratotrochus platypus Moseley, 1876, p. 554.
Stephanotrochus platypus; Moseley, 1881, p. 154, pl. 3, figs. 4a, 4b, 5a-5c.--Not S. platypus; Jourdan, 1895, pp. 19, 20 (is S. nobilis (Moseley, 1873)).
Not Stephanotrochus diadema var. platypus; Gravier, 1920, pp. 46, 47 (is S. moseleyanus (Sclater, 1886)).
Stephanocyathus sp. Squires and Ralph, 1965, pp. 262, 263, figs. 3, 4.

Stephanocyathus (S.) sp. Squires and Keyes, 1967, p. 24, pl. 2, figs. 11, 12.
Stephanocyathus platypus; Zibrowius, 1980, p. 97.

Description. Corallum free, bowl shaped, up to 75 mm in CD and 30 mm tall. Theca initially flat to slightly concave up to CD of 35-40 mm, then calicular edges turn upward rather abruptly and continue to grow at an angle of 60°-70° from horizontal. Costae not prominent on horizontal section, but C_1 and C_2 usually sharply ridged on upturned peripheral theca. Theca, except for C_1 and C_2, covered by very fine, low, rounded granules.

Septa hexamerally arranged in five cycles; S_6 common in larger coralla of up to 115 septa. S_1 extraordinarily exsert in form of rounded lobe projecting up to 16 mm beyond theca. S_2 also highly exsert; remaining septa barely exsert, only those flanking S_1 and S_2 rising higher than S_3. Calicular margin scalloped, apices corresponding to S_1 and S_2. S_1 extending to center of calice, there considerably thickened and fused into rudimentary columella. S_2 reaching almost to center and joining in fusion. S_3 falling just short of fusion and terminating in slightly lobed free end. S_4 slightly smaller than S_3 and also bearing small, broad paliform lobe, this lobe bending toward adjacent S_3 but rarely fusing with it. Where pairs of S_6 present, enclosed S_5 enlarged to almost size of S_4 and also bearing small lobe bending toward adjacent S_4. Normally, S_5 and all S_6 short, extending only one third of distance to center. S_3, S_4, and enlarged S_5 bearing broad, low paliform lobes not separated by notches; S_1 and S_2, however, without lobes and usually uniformly concave below level of theca. Septa straight with smooth inner edges. Septal granules low, blunt, and arranged in poorly defined lines.

Discussion. There is little doubt that Squires and Ralph's [1965] Stephanocyathus sp. and Moseley's S. platypus are identical. Moseley's [1876] original description was based on two small specimens with flat bases and calicular edges that had not yet turned upward; Squires and Ralph's very large specimen had an originally flat base that had subsequently become deeply bowl shaped. The three Eltanin specimens confirm the continuity of the ontogeny. S. diadema (Moseley, 1876) also has a flat base as a juvenile, which, like that of S. platypus, curves upward with greater size. Likewise, the base of S. laevifundus Cairns, 1979, is usually flat but is sometimes gently bowl shaped.

S. platypus is most similar to S. moseleyanus (Sclater, 1886) from the northeast Atlantic, especially in shape and septal exsertness. The latter is distinguished by its papillose columella, P_1 and P_2, and septal junctions near the columella.

Material. Eltanin sta. 1718 (2), USNM 47522; sta. 1818 (1), USNM 47423. Syntypes.

Types. Two syntypes of S. platypus, collected at Challenger station 164, are deposited at the British Museum (1880.11.25.57). Type-locality: 34°13'S, 151°38'E (off Sydney, Australia); 750 m.

Distribution. Known from only four records from off Sydney, Australia; off New Zealand; and from a seamount (Eltanin station 1718) east of New Zealand (Map 5). Depth range: 622-913 m. Like the pre-

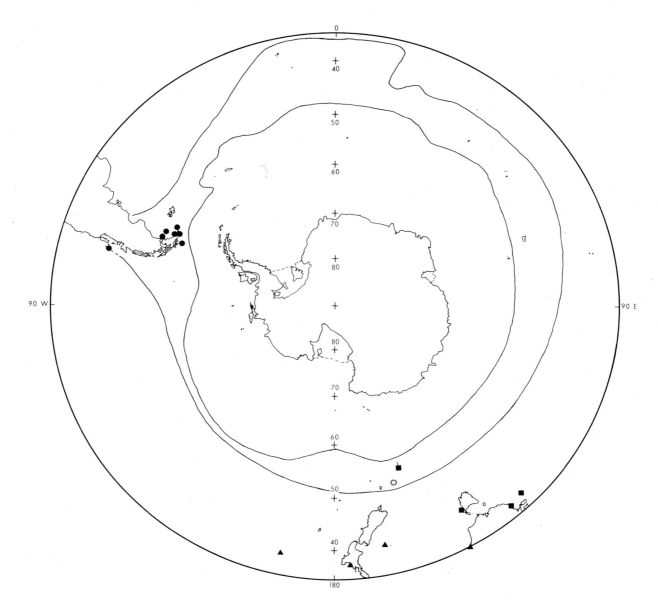

Map 5. Distribution of Sphenotrochus gardineri (solid circles), Aulocyathus recidivus (solid squares), Stephanocyathus platypus (solid triangles), and Lophelia prolifera (open circle).

vious species, it does not occur in the Subantarctic region as defined by Hedgpeth [1969] but is included here because of its proximity to the Subantarctic region.

Genus Aulocyathus Marenzeller, 1904

Diagnosis. Solitary, ceratoid, free. Longitudinal parricidal budding common. Columella trabecular. Type-species: Aulocyathus juvenescens Marenzeller, 1904a, by monotypy.

16. Aulocyathus recidivus (Dennant, 1906) n. comb.
Plate 7, figs. 7-9; Plate 8, fig. 1

Ceratotrochus recidivus Dennant, 1906, pp. 159, 160, pl. 6, figs. 2a-2c.--Squires, 1961, p. 18; 1969, p. 16.--Zibrowius, 1980, p. 107.
Ceratotrochsu [sic] (Conotrochus) typus; Wells,

1958, pp. 265, 266, pl. 1, figs. 14, 15.
? Paracyathus conceptus; Squires and Keyes, 1967, p. 23 (part: C-648, pl. 2, figs. 7, 8).

Description. The following description is based primarily on the largest specimen from NZOI station C-734. Corallum ceratoid, straight with round calice. Corallum 10.3 mm in CD and 17.8 mm tall. Most specimens originally attached to internal surface of fragment of parent specimen, from which they probably asexually budded. Some specimens originating from calice of unbroken parent specimen. Theca glossy, granular, sometimes marked by shallow striae. Costal granulation often indistinct and irregular. Calice usually round but may be elliptical.

Septa hexamerally arranged in four cycles. S_1 larger than S_2, these only slightly larger than S_3. S_4 smallest septa and never present as full cycle. Hexameral symmetry of younger specimens often not present; seven, eight, or nine groups of 4 or 6 septa often found. Large specimen with 40 septa includes two complete systems and four systems missing one pair of S_4 each. Septa not exsert with straight, vertical inner edges. Septal granules small and blunt, uniformly distributed.

Fossa deep. Columella composed of 11 individual, irregularly shaped rods.

Remarks. It is uncertain whether the corallum splits before the bud forms or whether the growing bud causes the corallum to fracture. Dennant [1906] implies the latter, whereas Marenzeller [1904a] implies the former for a related species, A. juvenescens.

Discussion. There are three other nominal species of Aulocyathus. A. recidivus differs from the two species known from off Japan, A. mactricidum (Kent, 1871) and A. conotrochoides (Yabe and Eguchi, 1932), by having a large, distinct columella. It differs from A. juvenescens Marenzeller, 1904 (off East Africa, 400-463), in being less slender and having fewer septa at a corresponding calicular diameter. Wells's [1958] specimen measures 11.6 mm in CD, is 25.6 mm tall, and has 60 septa. Squires's [1969] reference to C. recidivus from the Macquarie Ridge was undoubtedly from NZOI station C-734.

Material. NZOI sta. C-734 (9), USNM 47524. Golden Hind sta. 35 (1), MCZ. Specimen of Wells [1958] identified as C. (C.) typus from Discovery sta. 115, South Australian Museum H 51.

Types. The 'numerous' syntypes of C. recidivus are not at the Australian Museum [Zibrowius, 1980] and have not been traced further.

Distribution. Off southeastern Australia; off Tasmania; Macquarie Ridge (Map 5.) Depth range: 128-732 m.

Subfamily TURBINOLIINAE Milne Edwards and Haime, 1848
Genus Sphenotrochus Milne Edwards and Haime, 1848

Diagnosis. Solitary, cuneiform, free; corallum small. Theca imperforate. Costae deeply ridged or reduced to aligned granules. Columella lamellar or papillose. Type-species: Turbinola crispa Lamarck, 1816, by subsequent designation [Milne Edwards and Haime, 1850].

17. Sphenotrochus gardineri Squires, 1961
Plate 8, figs. 2-8

Sphenotrochus intermedius; Gardiner, 1939, p. 333 (part: Discovery sta. 388).
Sphenotrochus gardineri Squires, 1961, pp. 26-28, 29, text figs. 6-8; 1969, p. 17, pl. 6, map 1.--Cairns, 1979, p. 206.

Description. Corallum attached in younger stage. Base usually evenly rounded, but lateral edges of some specimens may form an apical angle of 50°-60°. Corallum cuneiform, highly compressed, with range of GCD/LCD of 1.5-2.2. Largest specimen known (Vema sta. 15-109) 9.3 x 4.4 mm in CD and 10.1 mm tall. Prominent, vertical costae correspond to all septa, with infrequent branching. The two principal (directive) costae, occurring on lateral edges, continuous to base, as are some costae on lateral faces; however, most costae originating at intercostal grooves flanking principal costae (Plate 8, fig. 3). Costae separated by deep, wide intercostal grooves. Row of low, rounded granules on each costa, sometimes becoming two granules wide near calice.

Septa hexamerally arranged in four cycles. S_1, S_2, and S_3 equal in size and exsertness, except for the 2 principal septa, these aligned with columella and considerably larger than other four S_1. S_4 considerably smaller; S_5 sometimes present in half systems adjacent to principals. Coralla with less than 48 septa, 10 or 11 half systems, common. S_{1-3} having coarsely dentate upper and inner edges, these merging with columella. Inner edges of S_4 finely dentate. Septal granules large and blunt, arranged in distinct lines originating from an axis of divergence well inside theca.

Fossa shallow. Columella composed of five to eight linearly arranged rods, these fused basally; never lamellar.

Remarks. Although adult coralla are always free, young specimens are initially attached to a substrate, such as sand particles of 0.1-0.2 mm in diameter. At a GCD of 2.5 mm, coralla have usually attained a full third cycle of septa, but costae and columella have not yet developed. A glossy epitheca is usually present. At a greater size, costae begin to develop and eventually overgrow the epitheca and original attachment, including the sand particle, at which point the coral becomes free. Coralla may also bud asexually from fragments of a parent specimen.

Discussion. S. gardineri is distinguished from other Recent Sphenotrochus, all of which are fairly localized in distribution. The following species are distinguished on the basis of their lamellar columella: S. hancocki Durham and Barnard, 1952 (Galapagos Islands; off California); S. auritus Pourtales, 1874 (tropical western Atlantic); S. gilchristi Gardiner, 1904; S. aurantiacus Marenzeller, 1904 (both off South Africa); and S. excavatus T. Woods, 1878 (off Australia). Most of these species also have only three cycles of septa. S. andrewianus Milne Edwards and Haime, 1848 (northeast Atlantic), has a lamellar or papillose columella but only three cycles of septa. Finally, S. ralphae Squires, 1964 (off New Zealand) [Squires, 1964b], has a similar columella, but its corallum is smaller and narrower and has fewer septa.

Material. Eltanin sta. 980 (2), USNM 47430. Hero sta. 712-656 (5), USNM 47432; sta. 715-694 (1), USNM 47434; sta. 715-902 (5), USNM 47433; sta. 715-903 (5), USNM 47428. Vema sta. 14-14 (4), USNM 47435; sta. 14-16 (1), USNM 47431; sta. 15-109 (2), USNM 47429; sta. 17-11 (2), USNM 47427. Specimens of Gardiner [1939] from Discovery sta. 388 (2), MCZ; specimens of Squires [1961] from Vema sta. 14-14 (5), AMNH. Holotype.

Types. The holotype is deposited at the American Museum of Natural History (3367). It was collected at Vema station 14-14. Type-locality: 54°23'S, 62°25'W (Burdwood Bank); 75 m.

Distribution. Endemic to Magellanic region from Tierra del Fuego to Chiloe, Chile (Map 5). Depth range: 9-403 m.

Plate 8. Aulocyathus, Sphenotrochus, and Desmophyllum

1. Aulocyathus recidivus (Dennant): USNM 47524, NZOI sta. C-734, H = 20.1 mm.
2-8. Sphenotrochus gardineri Squires: 2-4, USNM 47429, Vema sta. 15-109, H = 10.6 mm, GCD = 9.7 mm; 5, 8, USNM 47428, Hero sta. 715-903, H = 3.5 mm, 4.4 mm, young stages with poorly developed costae; 6, 7, USNM 47432, Hero sta. 712-656, H = 3.5 mm, 3.6 mm, young attached stages showing budding from septal fragments.
9-12. Desmophyllum cristagalli Milne Edwards and Haime: 9-11, USNM 36367, Albatross sta. 2785, series of 3 specimens from same station illustrating formae cristagalli (GCD = 64.0 mm), capense (GCD = 50.4 mm), and ingens (GCD = 59.2 mm), respectively; 12, specimen identified as D. capense by Gardiner [1939], BM 1939.7.20.220, WS sta. 99, GCD = 66.4 mm.

Plate 9. Desmophyllum, Solenosmilia, Lophelia, and Goniocorella

1-3. Desmophyllum cristagalli Milne Edwards and Haime: 1, specimen identified as D. capense by Gardiner [1939], BM 1939.7.20.220, WS sta. 99, x1.7; 2, USNM 36367, Albatross 2785, x0.52, long slender corallum; 3, specimen from same lot illustrating pseudocoloniality produced by settlement of successive solitary coralla, x0.33.

4, 5. Solenosmilia variabilis Duncan: 4, USNM 47426, Eltanin sta. 1346, x1.4, illustrating dichotomous extratentacular budding; 5, specimen from same lot, x0.54.

6. Lophelia prolifera (Pallas): USNM 47525, Eltanin sta. 1411, x3.2.

7-9. Goniocorella dumosa (Alcock): 7, USNM 47505, NZOI sta. D-175, x3.2, endothecal dissepiments; 8, 9, specimens from same lot, x0.46 and x9.8, respectively.

Subfamily DESMOPHYLLIINAE Vaughan and Wells, 1943
Genus *Desmophyllum* Ehrenberg, 1834

Diagnosis. Solitary, trochoid, fixed. No pali. Columella absent or very small. Sparse endothecal dissepiments. Type-species: *Desmophyllum dianthus* Ehrenberg, 1834, by subsequent designation [Milne Edwards and Haime, 1850].

18. *Desmophyllum cristagalli* Milne Edwards and Haime, 1848
Plate 8, figs. 9-12; Plate 9, figs. 1-3

The synonymy is complete for southern records only.
Desmophyllum cristagalli Milne Edwards and Haime, 1848, p. 253, pl. 7, figs. 10, 10a.--von Marenzeller, 1904a, pp. 267, 268, pl. 15, figs. 2a, 2b.--Gardiner, 1929a, pp. 125, 126.--Hoffmeister, 1933, pp. 8, 9, pl. 2, figs. 1-4.--Ralph, 1948, p. 108, fig. 2 (bottom left).--Wells, 1958, p. 262.--Squires, 1958, p. 91; 1961, pp. 18, 19; 1965, p. 785; 1969, p. 17, pl. 6, map 1.--Ralph and Squires, 1962, pp. 9, 10, pl. 3, figs. 1-10.--Squires and Keyes, 1967, p. 25, pl. 3, figs. 12-14.--Zibrowius, 1974a, pp. 758-761, pl. 3, figs. 1-10; 1980, pp. 117-121, pl. 61, figs. A-O, pl. 62, figs. A-M.--Beurois, 1975, p. 46, photo 13.--Cairns, 1979, pp. 117-119, pl. 21, figs. 7, 8, pl. 22, fig. 8.
Desmophyllum ingens Moseley, 1881, pp. 160-162, pl. 4, figs. 1-6, pl. 5, figs. 1-4a.--Squires, 1969, p. 17, pl. 6, map 1.
Desmophyllum capense; Gardiner, 1939, pp. 329, 330.--Wells, 1958, p. 262.--Cairns, 1979, p. 206.
Desmophyllum capensis; Squires, 1961, p. 23, fig. 5.

Description. The typical form of *D. cristagalli* has been adequately described elsewhere [Cairns, 1979; Zibrowius, 1980]; only a brief diagnosis is given here. Corallum variable in shape, from cylindrical to ceratoid, often greatly flared. Firmly attached by thick pedicel. Up to 80 x 50 mm in CD but averaging about 45 x 35 mm. Theca thick, covered by low, fine, rounded granules; ridged costae sometimes corresponding to S_{1-3}. Septa hexamerally arranged in five cycles, rarely with additional S_6. Septa thick and widely spaced, about 6-12/cm. S_1 and S_2 equal in size and exsert; septa of remaining cycles progressively smaller and less exsert. Inner septal edges straight; septal faces smooth, covered by low, rounded granules. Fossa deep, endothecal dissepiments sometimes present in elongate specimens. Columella rare, usually absent.

Forma *ingens*: Usually larger than typical form and often producing pseudocolonial clumps of specimens. S_6 common, up to 192 septa. Septa usually thinner and more crowded, about 14-18/cm. (Although Moseley cited only five cycles of septa in his original description of *D. ingens*, his illustrations delineate a specimen with about 136 septa, and syntypes from *Challenger* station 307 have up to 184 septa.)

Forma *capense*: Similar to *ingens* in size and septal arrangement. Differing from it in possessing distinct and often deep longitudinal grooves in theca; grooves partly partitioning calice into numerous scalloped lobes. Lobes increasing perimeter of calice, thus allowing space for more septa, up to 324 in largest specimen examined.

Remarks. Squires [1965b] cited pseudocolonial

D. cristagalli as the framework coral for a deepwater (334 m) coral bank on the Campbell Plateau, New Zealand. The colonial coral *Goniocorella dumosa* is also associated with these banks. Judging by the quantity of *D. cristagalli* dredged off Chile, it may also form deepwater coral banks there at depths of 300-800 m. Large clumps of specimens, exhibiting four or five successive settlements, are common in the *Albatross* material (Plate 9, fig. 3). No associated colonial ahermatype was found, although *L. prolifera*, *M. oculata*, *E. profunda*, and *S. variabilis* are usually found on deepwater banks in the Atlantic. The only associated corals at the Chilean stations were *Caryophyllia* sp. and *Javania cailleti*. Approximately 80% of the coralla were dead when collected.

Discussion. Von Marenzeller [1904a], Hoffmeister [1933], and Zibrowius [1974a, 1980] considered *D. ingens* to be a junior synonym of *D. cristagalli*. Gardiner [1929a] considered *D. ingens* to be distinct, and Ralph and Squires [1962] and Squires [1969] were inconclusive. *D. capense* has generally been acknowledged as a distinct species [Gardiner, 1904, 1939; Squires, 1961; Zibrowius, 1974a, 1980]. On the basis of a large suite of approximately 690 specimens from *Albatross* stations 2781 and 2785 (very near the type-locality of *D. ingens*) I have synonymized both *D. ingens* and *D. capense* sensu Gardiner, 1939, with *D. cristagalli*. Although I have examined only one syntype of *D. capense* Gardiner, 1904, this species is probably also a synonym of *D. cristagalli*. The *Albatross* stations contain typical specimens of *D. cristagalli*, forma *ingens*, and forma *capense*, as well as intermediates in size and morphology between *cristagalli* and *ingens* and between *ingens* and *capense*. *D. cristagalli* is known to be an extremely variable species [Cairns, 1979; Zibrowius, 1980] and one with few diagnostic characters (e.g., no columella, pali, or budding). My synonymy is based on this capacity for variation and the continuous suite from the *Albatross* stations. Their retention as formae is an artificial separation and is retained here only to aid future revisers.

Material. Forma *cristagalli*: *Eltanin* sta. 255 (2), USNM 47396; sta. 1345 (4), USNM 47406; sta. 1346 (18), USNM 47407; sta. 1403 (4), USNM 45668; sta. 1411 (5), USNM 47394; sta. 1422 (3), USNM 47395; sta. 1605 (1), USNM 47400; sta. 1691 (3), USNM 47402; sta. 1718 (5), USNM 47293; sta. 1851 (12), USNM 47412. *Albatross* sta. 2782 (1), USNM 36352. NZOI sta. C-734 (1), USNM 47404; sta. C-618 (3), USNM 47411; sta. D-145 (2), USNM 53381; sta. D-149 (2), USNM 47401; sta. D-159 (2), USNM 47408; sta. D-160 (1), USNM 47410; sta. D-166 (1), USNM 47403; sta. D-175 (40), USNM 47413; sta. D-176 (9), USNM 53377. Forma *ingens*: *Eltanin* sta. 214 (1), USNM 47398; sta. 25-326 (9), USNM 47405; sta. 369 (1), USNM 47399; sta. 740 (3), USNM 45669; sta. 958 (1), USNM 47397; sta. 1536 (4), USNM 47409. *Vema* sta. 17-39 (1), USNM 47414. WH sta. 311/66 (2), ZIZM; sta. 361/66 (1), ZIZM. Forma *capense*: WH sta. 361/66 (1), ZIZM. Mixture of all three formae: *Albatross* sta. 2781 (180), USNM 19167; sta. 2785 (510), USNM 36367. Specimens (3) of *D. capense* [Gardiner, 1939], BM 1939.7.20.220-223; specimens listed by Cairns [1979]. Holotype of *D. cristagalli*; 3 syntypes of *D. ingens* (BM 1880.1$\overline{1}$.25.67); syntype of *D. capense* (MCZ).

Types. The holotype of *D. cristagalli* is deposited at the Muséum National d'Histoire Naturelle, Paris, and is illustrated by Cairns

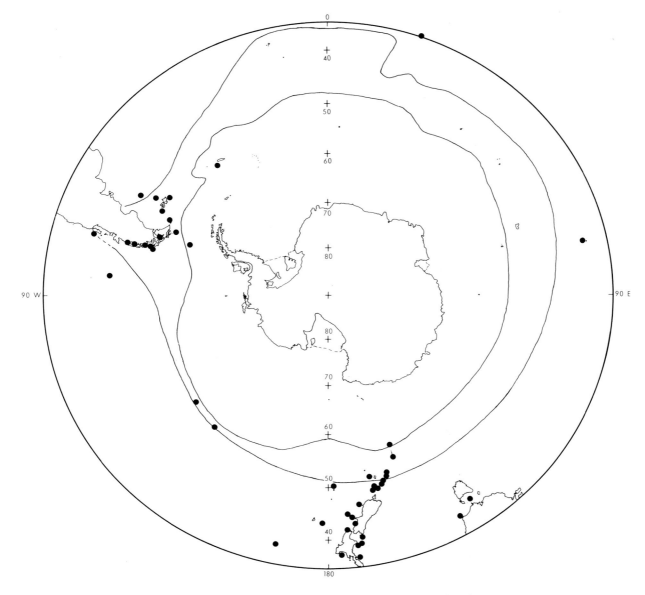

Map 6. Distribution of <u>Desmophyllum</u> <u>cristagalli</u>.

[1979]. Type-locality: Gulf of Gascony; depth unknown. The syntypes of <u>D</u>. <u>ingens</u> are deposited at the British Museum. Type-locality: fjords of southern Chile from 48°27'S to 52°45'S; 256-631 m. The syntypes of <u>D</u>. <u>capense</u> are deposited at the British Museum and the Museum of Comparative Zoology (3885). Type-locality: Cape Hangklip, South Africa; 81 m.

<u>Distribution</u>.   One of the few cosmopolitan species of Scleractinia, widely distributed in the Atlantic, Pacific, and Indian oceans. In southern seas, found off southern South America, Falkland Islands, South Georgia, ? South Africa, Île Saint-Paul and Île Amsterdam, southeastern Australia, New Zealand, Auckland Island, Macquarie Island, Hjort Seamount, and several Subantarctic seamounts in South Pacific. Not present off continental Antarctica. Forma <u>ingens</u> restricted to off southern tip of South America and off South Georgia. Forma <u>capense</u> known from off southern Chile, off Falkland Islands, and off South Africa

(Map 6). Worldwide depth range: 35-2460 m; Subantarctic records: 91-1463 m.

Genus <u>Lophelia</u> Milne Edwards and Haime, 1849

<u>Diagnosis</u>.   Colonial, forming large dendroid colonies by intratentacular budding.   Coenosteum dense.   Costae and columella poorly developed. Pali absent.   Sparse tabular endothecal dissepiments. Type-species: <u>Madrepora</u> <u>prolifera</u> Pallas, 1766, by subsequent designation [Milne Edwards and Haime, 1850].

19. <u>Lophelia prolifera</u> (Pallas, 1766)
Plate 9, fig. 6

<u>Discussion</u>.   A small branch fragment, containing only five damaged corallites, was collected at <u>Eltanin</u> station 1411 (USNM 47525) from a seamount on the north Macquarie Ridge just within the Subantarctic region. Full descriptions and synonymy

for this species can be found in the works by Cairns [1979] and Zibrowius [1980].

L. prolifera is a widely distributed species, found throughout the Atlantic Ocean, including off Tristan Island; off South Africa; off northern Madagascar; off Île Saint-Paul and Île Amsterdam; and probably off California. The Eltanin record is the southernmost report of L. prolifera and the second for the Subantarctic region, the other being Moseley's [1881] record off Tristan Island (Map 5). Worldwide bathymetric range: 60–2170 m.

Subfamily PARASMILIINAE Vaughan and Wells, 1943
Genus Solenosmilia Duncan, 1873

Diagnosis. Colonial, dendroid, or subplaceloid colonies formed by intratentacular budding. Stereome granular, costae sometimes corresponding to first cycle. Tabular endothecal dissepiments. Columella small. Type-species: Solenosmilia variabilis Duncan, 1873, by monotypy.

20. Solenosmilia variabilis Duncan, 1873
Plate 9, figs. 4, 5

The synonymy is complete for southern records only.
Solenosmilia variabilis Duncan, 1873, p. 328, pl. 42, figs. 11–18.--Moseley, 1881, p. 181, pl. 9, figs. 1–5.--von Marenzeller, 1904a, pp. 310, 311, pl. 15, figs. 4, 4a.--Hoffmeister 1933, p. 14, pl. 14, fig. 7.--Wells, 1958, p. 262.--Squires, 1969, pp. 16, 18, pl. 6, map 2.--Zibrowius, 1974a, pp. 768, 769; 1980, pp. 143–145, pl. 75, figs. A–N.--Cairns, 1979, pp. 136–138, pl. 26, figs. 2–4.

Description. This species has recently been described elsewhere by Cairns [1979] and Zibrowius [1980]; a brief description follows. Colonies bushy with frequent anastomosis of branches; intratentacular budding. Terminal branches and calices about 6 mm in diameter. Coenosteum smooth, granular, or costate; white or grayish. Septa hexamerally arranged but very irregular in development. S_4 sometimes present, but rarely as complete cycle. Septal granulation sometimes very prominent with slender granules as high as 3–4 times septal width. Tabular endothecal dissepiments. Columella usually absent but may be small and spongy.

Discussion. Because of its distinctive branching pattern, S. variabilis is easily distinguished from the four other colonial genera that occur in Subantarctic waters: Madrepora, Enallopsammia, Bathelia, and Goniocorella. Both S. variabilis and M. oculata were collected at Eltanin station 1081 (east of the South Orkney Islands), which is the southernmost record for colonial Scleractinia.

S. variabilis was taken in great quantity at Eltanin station 1346 (a seamount or ridge on the Heezen fracture zone of the Eltanin fracture zone system) in the South Pacific, indicating the possible presence of a deepwater coral bank. Other similar deepwater structures have been found in the North Atlantic and on the Campbell Plateau, south of New Zealand. The framework corals of the northeastern Atlantic banks are L. prolifera and Madrepora oculata (S. variabilis is present to a minor extent); those in the northeast Atlantic are Enallopsammia profunda and L. prolifera (again S. variabilis is present but not common); those on the Campbell Plateau are Goniocorella dumosa and

pseudocolonial Desmophyllum cristagalli. The coral composition of the South Pacific bank is apparently about 98% S. variabilis with a small amount of M. oculata. Other associated solitary corals are Desmophyllum cristagalli, Cyathoceras irregularis, and Caryophyllia sp. Other invertebrates found at the same station include Porifera, Hydroida, Stylasterina, Gorgonacea, Actiniaria, Nematoda, Bryozoa, Brachiopoda, Ophiuroidea, Asteroidea, Echinoidea, Holothuroidea, Pterobranchia, Polychaeta, Gastropoda, Polyplacophora, Bivalvia, Pycnogonida, and Crustacea.

Without a sediment sample, seismic profile, and photographic documentation it is difficult to be conclusive, but in all likelihood a deepwater coral bank exists in this area.

Material. Eltanin sta. 254, USNM 47423; sta. 1081, USNM 47422; sta. 1344, USNM 47424; sta. 1345, USNM 47425; sta. 1346, USNM 47426; sta. 1403, USNM 47419; sta. 1414, USNM 47420; sta. 1416, USNM 47664; sta. 1422, USNM 47421; Specimens listed by Cairns [1979], USNM. Syntypes.

Types. The syntypes of S. variabilis, collected on the second cruise of the Porcupine, are deposited at the British Museum. Type-locality: off southwestern Spain; 1190–2003 m.

Distribution. Widespread in the Atlantic and Indian oceans. Circumpolar in southern seas: off South Africa; off Prince Edward Island; off Île Saint-Paul; off southeastern Australia; Hjort Seamount; Macquarie Ridge; off New Zealand; seamounts in South Pacific and Drake Passage; east of South Orkney Islands; off Tristan Island (Map 7). Not found off continental Antarctica. Squires's [1969] record off Chile is unsubstantiated. Worldwide depth range: 220–2165 m; Subantarctic records: 500–1830 m.

Genus Goniocorella Yabe and Eguchi, 1932

Diagnosis. Colonial, extratentacular budding forming bushy, dendroid colonies. Branches anastomose and also joined by slender extensions of coenosteum. No columella or pali. Tabular endothecal dissepiments widely spaced. Type-species: Pourtalosmilia dumosa Alcock, 1902, by original designation.

21. Goniocorella dumosa (Alcock, 1902)
Plate 9, figs. 7-9; Plate 10, figs. 1, 2

Pourtalosmilia dumosa Alcock, 1902, pp. 36, 37, pl. 5, figs. 33, 33a.
Goniocorella dumosa; Yabe and Eguchi, 1932, pp. 389, 390; 1936, p. 167; 1941b, pp. 162, 163; 1943, pp. 495, 496, figs. 1, 2.--Squires, 1960, pp. 197, 198, pl. 33, figs. 1–4; 1964a, p. 11; 1965b, pp. 785–787; 1969, p. 17, pl. 6, map 2. --Ralph and Squires, 1962, p. 11, pl. 4, fig. 1. --Squires and Keyes, 1967, p. 25, pl. 3, figs. 15, 16, text fig. 4.--Eguchi, 1968, C-43, pl. C-9, figs. 11, 12.--Podoff, 1976, pp. 27, 28, pl. 1, figs. 5, 6.

Description. Colony bushy, achieved by irregular extratentacular budding often at right angles to parent branch. Parent branch continuing to grow after budding; each budded branch elongating and also producing buds. Strength of colony reinforced by numerous slender (2 mm in diameter) extensions of coenosteum, these uniting adjacent branches,

Plate 10. _Goniocorella_ and _Flabellum_

1, 2. _Goniocorella dumosa_ (Alcock): 1, USNM 47505, NZOI sta. D-175, x2.9, illustrating coenosteal processes; 2, specimen from same lot, CD = 4.1 mm,
illustrating dissepiment forming.

3-5. _Flabellum thouarsii_ Milne Edwards and Haime: 3, 4, syntype, MNHNP 1028, H =
21.8 mm, GCD = 24.9 mm; 5, specimen reported by Studer [1878], Museum für
Naturkunde, Berlin, number 1737, Gazelle sta. 54, CD = 22.7 x 15.1 mm.

6-11. Series of 3 species of _Flabellum_ of about the same GCD: 6, 7, _Flabellum_
thouarsii, USNM 47222, _Eltanin_ sta. 976, GCD = 21.2 mm; 8, 9, _Flabellum_
impensum, USNM 45629, EW sta. 9, GCD = 23.9 mm; 10, 11, _Flabellum_ _curvatum_,
USNM 47254, _Eltanin_ sta. 558, GCD = 22.6 mm.

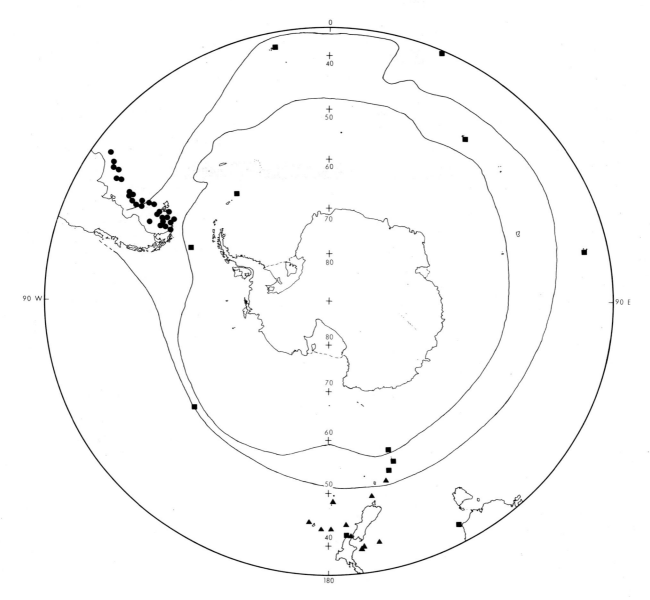

Map 7. Distribution of _Solenosmilia variabilis_ (solid squares), _Goniocorella dumosa_ (solid triangles), and _Flabellum thouarsii_ (solid circles).

sometimes in scalariform arrangement [Squires, 1960]. Branches cylindrical, often straight, 3-5 mm in diameter, each bearing round terminal calice. Colonies up to 1 m in diameter. Corallum light gray or white; polyps and tentacles orange. Coenosteum bearing low, rounded granules. Terminal corallites often with slightly ridged C_1 and C_2.

Septa hexamerally arranged in three cycles. S_1 very slightly exsert, with straight, vertical inner edges. Upper region of septa usually narrower than lower region; lower region almost reaching center of corallite. S_2 and S_3 progressively smaller; S_3 rudimentary, with dentate inner edges. Septal faces usually smooth with fine granulation, but sometimes covered with tall, pointed granules.

Fossa deep and vacuous. No columella or pali. Thin, tabular endothecal dissepiments occurring

every 2-10 mm, giving dried corallum a light weight.

Remarks. On the basis of a specimen attached to an underwater cable, Squires [1960] calculated the growth rate to be at least 1.67-2.94 mm/year in height. Squires [1965b] also suggested that _G. dumosa_ is the primary sediment-forming coral of a deepwater coral bank (coppice) on the Campbell Plateau, off New Zealand. Two other locations of _Goniocorella-Desmophyllum_ coppices are given by Squires [1965b].

Discussion. _Goniocorella_ is monotypic and distinguished from other southern colonial corals by its distinctive branching pattern, coenosteal extensions, and lack of columella.

Material. Eltanin sta. 1816, USNM 47509; sta. 1848, USNM 47667. NZOI sta. A-706, USNM 47511; sta. B-319, USNM 47506; sta. C-410, USNM 47510; sta. C-618, USNM 47507; sta. C-633, USNM 47508;

sta. D-145, USNM 53382; sta. D-175, USNM 47505. Some of these records first reported by Ralph and Squires [1960], Squires [1965b], and Squires and Keyes [1967].

Types. The syntypes of G. dumosa are at the Indian Museum (Calcutta). Type-locality: Banda Sea; 469-487 m.

Distribution. Off Japan; Banda Sea; off Norfolk Island; off Bounty Islands; off New Zealand; Chatham Rise; Campbell Plateau (Map 7). Depth range: 100-638 m.

Superfamily FLABELLICAE Bourne, 1905
Family FLABELLIDAE Bourne, 1905
Genus Flabellum Lesson, 1831

Diagnosis. Solitary, cuneiform to compressed turbinate, free. Wall epithecal. Base not thickened by stereome; no roots. Calicular edge jagged or entire. Pali absent. Columella rudimentary or absent. Type-species: Flabellum pavoninum Lesson, 1831, by subsequent designation [Milne Edwards and Haime, 1850].

22. Flabellum thouarsii Milne Edwards and Haime, 1848
Plate 10, figs. 3-7

Flabellum thouarsii Milne Edwards and Haime, 1848, p. 265, pl. 8, fig. 5.--Studer, 1878, p. 630.--Not F. thouarsii; Gravier, 1914b, pp. 125-128 (is F. impensum Squires, 1962).--Not F. thouarsii; Wells, 1958, p. 268 (is F. impensum Squires, 1962, and F. flexuosum, n. sp.--Squires, 1961, pp. 29-38, figs. 5, 14-19, 21, 23, 27; 1962b, pp. 14, 18; 1969, p. 18, pl. 6, map 4.--Keller, 1974, pp. 200-203, pl. 1, figs. 1-4, pl. 2, figs. 1-7, pl. 3, figs. 1-4, pl. 4, figs. 1-7.--Cairns, 1979, p. 206.
Flabellum thouarsi; Milne Edwards and Haime, 1857, pp. 89, 90.
Flabellum curvatum; Gardiner, 1939, pp. 327, 328, pl. 20, figs. 1, 2 (part: all stations but Discovery sta. 182, WS sta. 839).--Squires, 1961, p. 38 (part: Vema sta. 14-18); 1962a, pp. 1-11, figs. 1-3.

Description. Corallum ceratoid to flabellate, attached only in young stage. Pedicel short, about 2.5-3.2 mm in diameter and usually worn away in older specimens. Pedicel expanding into straight corallum with angle of lateral edges varying from 30° (ceratoid) to 90° (flabellate); inclination of lateral faces rarely exceeding 20°. Largest specimen examined 33 x 22 mm in CD and 34 mm tall; an average-size specimen, however, 21 x 14 mm in CD and 23 mm tall. Theca usually worn or encrusted with calcareous invertebrates and ploychaete sand tubes. Calice elliptical, with GCD/LCD ratios of 1.3-1.7; calicular profile arched.

Septa hexamerally arranged in five cycles; however, only largest specimens with 96 septa. Most specimens with 20-22 major septa (S_{1-3}) enclosing triads of septa, or 80-88 total septa. Ten or eleven $S_{1,2}$ equal in size and reaching center of fossa, sometimes fusing with those on opposite side. Ten or eleven S_3 smaller, 20-22 S_4 one third to one fourth of size of S_3, and S_5 rudimentary. Septa not exsert; each larger septum having shallow, concave notch near calicular edge, but never dentate. Inner edges of larger septa

straight to slightly sinuous and thickened deeper in fossa. Septal granulation coarse and irregular, granules sometimes arranged in rows parallel to septal edge.

Fossa relatively shallow. Rudimentary columella formed by fusion of thickened inner edges of S_1 and S_2 and additional irregular extensions from these septa. Internal stereome present in older specimens.

Remarks. Judging from substrate of attachment and attached worm tubes, this species seems to inhabit bottoms composed of small pebbles and coarse sand. Squires [1962a] discusses planulation, size of larva, and lack of periodicity for this species. Specimens from the northern range seem to attain a larger size and are therefore more easily confused with F. curvatum.

Discussion. As Squires [1961, p. 31] implied, Gardiner [1939] was probably not aware of F. thouarsii; most of his records of F. curvatum are average- to large-size specimens of F. thouarsii from its characteristic depth range, and there is one record each of F. impensum and F. curvatum. Authors that have relied on Gardiner's identifications [e.g., Wells, 1958; Keller, 1974] have been misled. Although Squires [1961] differentiated F. thouarsii from F. curvatum, he included one suite of uncharacteristically large specimens of F. thouarsii as F. curvatum (Vema station 14-18). Furthermore, his paper on the larvae of F. curvatum [Squires, 1962a] is also based on large F. thouarsii. Keller [1974] recognized two forms of F. thouarsii, one similar to Squires's [1961] F. thouarsii and the other similar to his F. curvatum. She did not agree with Squires's [1961] reasons for separating the two species and therefore synonymized them. Keller's specimens all seem to be typical F. thouarsii. Ironically, her next species account, which describes F. antarcticum, is typical F. curvatum.

Comparisons of F. thouarsii to the closely related F. curvatum and F. areum are discussed with those species.

Material. Eltanin sta. 217 (1), USNM 45650; sta. 370 (22), USNM 45664; sta. 974 (232), USNM 45649; sta. 976 (184), USNM 47222; sta. 977 (309), USNM 45648. Edisto sta. 49 (1), USNM 45643; sta. 50 (1), USNM 47218. Vema sta. 14-6 (10), AMNH; sta. 14-16 (18), USNM 45654, and (13), AMNH; sta. 14-18 (1), USNM 45655; sta. 15-PD3 (6), AMNH; sta. 15-99 (14), USNM 47212, and (1), AMNH; sta. 15-102 (135), USNM 47220; sta. 15-103 (2), USNM 45651, and (3), AMNH; sta. 15-108 (6), USNM 47216, and (6), AMNH; sta. 15-110 (15), USNM 47221, and (19), AMNH; sta. 16-39 (8), USNM 47214, and (100), AMNH; sta. 17-74 (19), USNM 45617, and (8), AMNH; sta. 17-76 (40), USNM 47211, and (81), AMNH; sta. 17-88 (10), USNM 45618, and (10), AMNH; sta. 17-89 (13), USNM 47217, and (21), AMNH; sta. 17-90 (1), AMNH; sta. 17-97 (100), AMNH; sta. 18-13 (32), USNM 47219; sta. 18-14 (3), USNM 47215; sta. 18-16 (4), USNM 47213. BR sta. 25149 (2), USNM 47210. Specimens (2) of Studer [1878], Museum für Naturkunde, Berlin (number 1737); some Vema records first reported by Squires [1961, 1962a, b] as F. thouarsii and F. curvatum; some specimens identified as F. curvatum by Gardiner [1939]: WS sta. 76 (1), BM 1939.7.20.169; sta. 216 (19), BM 1939.7.20.252-264; sta. 244 (11), BM 1939.7.20.154-164; sta. 247 (1), BM 1939.7.20.170; sta. 792 (1), BM 1939.7.20.177 and Discovery sta. 652 (7), MCZ 3589. Syntypes.

Types. Milne Edwards and Haime's original description gave measurements for only one specimen, implying a holotype. However, at the Muséum National d'Histoire Naturelle, Paris, there are four specimens labeled F. thouarsi from the 'Îles Malouines' in the Milne Edwards Collection, two numbered 1028 and two numbered 1029. Three are F. thouarsii; a fourth from lot 1029 appears to be an Indo-Pacific hermatype of similar shape. One of the specimens from lot 1028 has measurements similar to those of the specimen in the original description and may be the holotype. Gravier [1914b, pp. 127, 128] reported two types (syntypes ?) of F. thouarsii at the Muséum National d'Histoire Naturelle and two additional specimens identified as this species, one from 'Îles Malouines' and the other 'trouvé dans une éponge.' This may explain the presence of four identified specimens of F. thouarsii but does not help in determining the type. Because of this uncertainty a lectotype is not chosen. Type-locality: Malouine Islands (Falkland Islands).

Distribution. Off southeastern South America from Rio de la Plata, Uruguay, to Cape Horn; off Falkand Islands. Squires's [1969] record from the Scotia Ridge undocumented (Map 7). Depth range: 71-305 m.

23. Flabellum areum n. sp.
Plate 11, figs. 1-5

Description. Corallum trochoid to turbinate, sometimes campanulate; usually straight. Weakly attached by short pedicel 3.2-5.1 mm in diameter; pedicel often detached from substrate (usually a small pebble), allowing base and pedicel to erode gradually. Largest specimen 27.7 x 20.0 mm in CD and 21.2 mm tall. Theca usually worn, bearing thin, longitudinal striae corresponding to all septa, characteristic of most flabellids. Calice elliptical, with GCD/LCD ratio usually between 1.3 and 1.4; lateral edges rounded. Calicular margin entire, calicular profile straight to slightly arched.

Septa hexamerally arranged in five cycles; however, fifth cycle never complete. One large specimen with 86 septa; holotype (adult specimen) with 80 septa. S_1 and S_2 equal in size and meeting in center of calice. S_1 and S_2 not exsert and forming near calicular edge shallow, concave notch; notch may or may not be dentate. Upper thecal edge forming lip rising slightly above upper septal insertions. Size of S_3 depending on presence of paired S_5 in half system. If S_5 present (which is more common in half systems adjacent to lateral edges), then S_3 one half to three fourths of size of $S_{1,2}$; when S_5 absent, S_3 about one third of size of $S_{1,2}$. S_4 and S_5 progressively smaller, S_4 extending only about halfway to base. Inner edges of septa usually straight but may be slightly sinuous. Septal granules low to moderately tall and arranged in lines parallel, or rows almost perpendicular, to septal edge, the latter corresponding to fine sinuosity of septal edge.

Fossa appearing spacious because of widely spaced septa and campanulate corallum shape. Lower, inner edges of S_1 and S_2 greatly thickened and fused in center, forming rudimentary columella. Theca within calice often increased substantially by deposits of stereome, sometimes obscuring smaller septa. Stereome also filling in base of more elongate coralla.

Remarks. The maximum size of this species probably lies between a GCD of 25-30 mm, judging from the extreme development of stereome and worn pedicels of specimens in this size range.

Discussion. F. areum is most similar to F. thouarsii, particularly in size, septal arrangement, and geographic distribution. It can be distinguished by its deeper fossa, larger pedicel diameter, lesser number of septa per centimeter (about 12 for F. areum, 18 for F. thouarsii), and much deeper bathymetric range.

Etymology. The specific name areum (Latin: open or vacant space) refers to the spacious fossa.

Material. Vema sta. 15-132 (25), AMNH; sta. 17-57 (15), AMNH; sta. 17-61 (3), USNM 47913. Types.

Types. The holotype, collected at Eltanin station 1592, is deposited at the United States National Museum (47167). Three paratypes from Eltanin station 1592 (number 47168), 14 from Eltanin station 973 (number 45639), 8 from Vema station 17-57 (number 47169), and 15 from Vema station 15-132 (number 45616) are deposited at the United States National Museum. Type-locality: 54°43'S, 55°30'W (Scotia Ridge east of Burdwood Bank); 1647-2044 m.

Distribution. Off Mar del Plata, Argentina; off Cape Horn, Tierra del Fuego; Scotia Ridge southeast of Falkland Islands (Map 8). Unconfirmed depth range: 1647-2229 m.

24. Flabellum curvatum Moseley, 1881
Plate 10, figs. 10, 11; Plate 11, figs. 6-9

Flabellum curvatum Moseley, 1881, pp. 174, 175, pl. 6, figs. 3a-3d.--Gardiner, 1939, pp. 327, 328 (part: WS sta. 839 only).--Squires, 1961 (part: not Vema sta. 14-18), pp. 7, 9, 29, 38, 39, figs. 5, 11-13, 20, 22, 30.--Not F. curvatum; Squires, 1962a, pp. 1-11, figs. 1-3 (is F. thouarsii Milne Edwards and Haine, 1848).--Squires, 1962b, p. 14; 1964a, p. 13; 1964c, pl. 1, fig. 1; 1969, p. 18, pl. 6, map 3.--Cairns, 1979, p. 206.
Flabellum antarcticum; Keller, 1974, pp. 203-205, pl. 5, figs. 1-7.

Description. Corallum ceratoid to trochoid, rarely attached above GCD of 12 mm. Pedicel long, slender (2.5-3.3 mm in diameter), and usually bent; base of pedicel often worn to a point in older specimens. Corallum usually curved, enlarging gradually from pedicel to calice. Largest specimen examined 44 x 30 mm in CD and 47 mm tall. Theca usually worn or encrusted with bryozoans, serpulids, or other corals; sometimes thin, incised costal lines present, one corresponding to each septum. Calice elliptical, with GCD/LCD ratios of 1.4-1.6; calicular profile arched.

Septa hexamerally arranged in five cycles. S_1 and S_2 equal in size and usually slightly larger than S_3. S_4 about half size of S_3; S_5 rudimentary and sometimes fenestrate because of weakly calcified trabeculae. Septa not exsert; upper edge of each larger septum forming shallow, concave notch near calicular edge. This notch often dentate but not always. Inner edges of larger septa straight to slightly sinuous and thickened lower in fossa. This thickening, along with loose fusion of irregular processes from lower inner edges of larger septa, forming rudimentary columella. Septal granulation variable, ranging

Plate 11. *Flabellum*

1-5.   *Flabellum areum* n. sp.: 1-3, holotype, USNM 47167, *Eltanin* sta. 1592, H = 21.9 mm, GCD = 25.8 mm; 4, paratype, USNM 45639, *Eltanin* sta. 973, H = 32.5 mm, corallum broken, revealing columella; 5, specimen (paratype) from same lot, H = 21.7 mm.

6-9.   *Flabellum curvatum* Moseley: 6, 7, syntype, BM 1974.8.510, *Challenger* sta. 320, GCD = 40.7 mm; 8, syntype, BM 1880.11.25.85, *Challenger* sta. 320, GCD = 29.7 mm; 9, USNM 47253, *Eltanin* sta. 340, x3.2, illustrating dentate septal notch.

10.   *Flabellum impensum* Squires: USNM 45666, EW sta. 37, H = 65.1 mm.

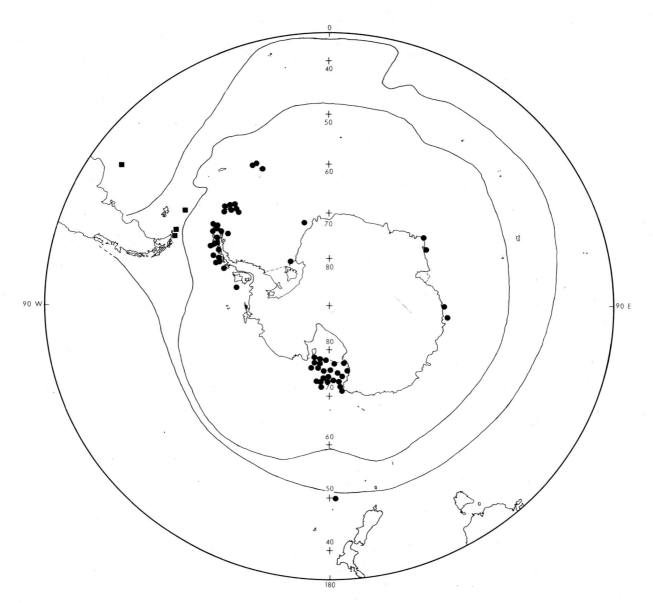

Map 8. Distribution of <u>Flabellum</u> <u>impensum</u> (solid circles), and <u>Flabellum</u> <u>areum</u> (solid squares).

from low and rounded to high, slender, pointed granules, usually arranged in rows subparallel to septal edges. Fossa elongate and relatively shallow. Stereome present in elongate specimens.

<u>Remarks</u>. F. <u>curvatum</u> usually settles on pebbles 3-4 times the diameter of its original attachment. Its bent pedicel and curved corallum probably reflect a reorientation of the polyp after it detaches from its substrate or when it becomes so heavy that it topples sideways. It also attaches to coralla of its own species, echinoid spines, and branching bryozoans. In turn, the theca of the living coral provides a substrate for numerous species of Bryozoa, serpulid polychaetes, barnacles, hydrocorals, and other scleractinians.

F. <u>curvatum</u> probably occurs in fairly high density off East Falkland island, indicated by the recovery of over 2500 specimens from a 68-min trawl.

<u>Discussion</u>. As is indicated by the synonymy, F. <u>curvatum</u> has often been confused with F. <u>thouarsii</u>, a closely related species. In fact, both Wells [1958] and Keller [1974] have synonymized these species. After thorough reexamination of this species complex I find that F. <u>curvatum</u> can be distinguished by a combination of the following characters: (1) the pedicel is usually bent and the corallum is usually curved; (2) the pedicel is longer; (3) the maximum size of the corallum is larger; (4) the septal notch is sometimes dentate; (5) the S_4 are relatively larger than those of F. <u>thouarsii</u>; and (6) the fossa is usually shallower, sometimes partially occupied by a crispate columella. Characteristics of attachment and pedicel diameter mentioned by Squires [1961] are of no diagnostic value. Furthermore, although their depth ranges overlap, F. <u>curvatum</u> is usually found deeper than F. <u>thouarsii</u>.

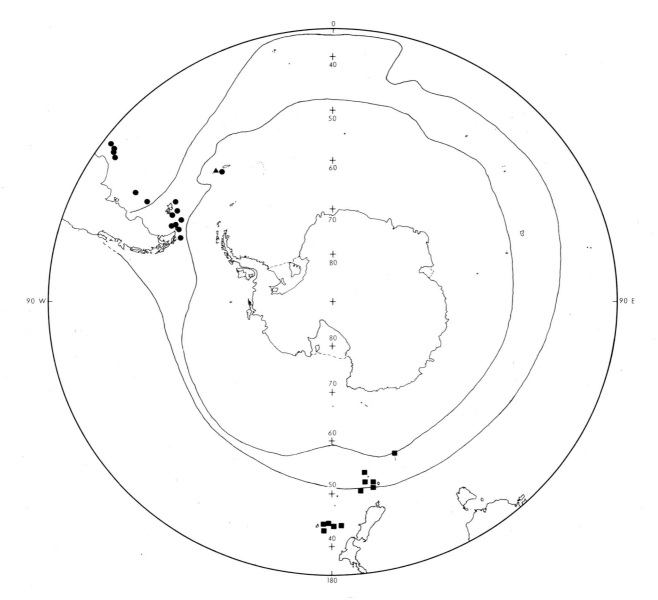

Map 9. Distribution of <u>Flabellum</u> <u>curvatum</u> (solid circles), <u>Flabellum</u> <u>knoxi</u> (solid squares), and <u>Flabellum</u> <u>gardineri</u> (solid triangle).

<u>F</u>. <u>curvatum</u> is distinguished from <u>F</u>. <u>impensum</u> by its coarser septal granulation, smaller PD, usually curved corallum, shallower fossa, and geographical distribution.

<u>Material</u>. <u>Eltanin</u> sta. 339 (73), USNM 47238; sta. 340 (139), USNM 47253; sta. 556 (8), USNM 47243; sta. 558 (about 2500), USNM 47254; sta. 740 (6), USNM 47239; sta. 1536 (11), USNM 47242. <u>Hero</u> sta. 715-875 (126), USNM 47251; sta. 715-885 (1), USNM 47245; sta. 715-895 (35), USNM 47252. <u>Vema</u> sta. 14-12 (4), USNM 45652, and (8), AMNH; sta. 15-PD3 (5), USNM 47240; sta. 15-PD4 (39), USNM 47250; sta. 15-PD9 (3), USNM 47241; sta. 15-PD10 (5), USNM 45626, and (11), AMNH; sta. 17-59 (2), USNM 45653; sta. 17-100 (24), USNM 45621, and (37), AMNH; sta. 17-101 (9), USNM 45624, and (32), AMNH; sta. 18-8 (7), USNM 45620; sta.. 18-12 (16), USNM 45622. <u>Calypso</u> sta. 171 (2), USNM 47246; sta. 172 (4), SME. BR sta.. 25149 (5), USNM 47244. Specimens of Squires [1961, 1962a], USNM; specimens of Gardiner

[1939] from WS sta. 839, BM 1939.7.20.129. Two syntypes (BM 1880.11.25.85 and 1974.8.5.10).

<u>Types</u>. The eight syntypes of <u>F</u>. <u>curvatum</u> are deposited at the British Museum. Type-locality: 37°17'S, 53°52'W (off Río de la Plata, Uruguay); 1097 m.

<u>Distribution</u>. Off southeastern South America from Río de la Plata, Uruguay, to Cape Horn; Burdwood Bank; off Falkland Islands; off South Georgia (Map 9). Depth range: 115-1137 m; however, most common between 400 and 800 m.

25. <u>Flabellum</u> <u>impensum</u> Squires, 1962
Plate 10, figs. 8, 9; Plate 11, fig. 10;
Plate 12, figs. 1-8

<u>Flabellum</u> <u>inconstans</u>; Pax, 1910, pp. 66-72, pl. 11, figs. 3-9, pl. 12, figs. 1-6.
<u>Flabellum</u> <u>thouarsii</u>; Gravier, 1914b, pp. 125-128, pl. 1, figs. 5, 6.--Wells, 1958, p. 268 (part :

Plate 12. Flabellum

1-8.   Flabellum impensum Squires: 1, 2, USNM 45629, EW sta. 9, H = 78.5 mm, GCD = 116.0 mm; 3, USNM 47356, Eltanin sta. 1878, H = 10.1 mm; 4, USNM 47368, Eltanin sta. 1084, x0.51, muricid gastropod egg cases; 5, USNM 45637, Eltanin sta. 437, x0.92, volutid gastropod egg case; 6, USNM 45635, Eltanin sta. 426, x1.7, broken corallum revealing columella; 7, USNM 45666, EW sta. 37, x1.9, dentate and fenestrate septal edges; 8, specimen identified as F. transversale by Thomson and Rennet [1931], Australian Museum G 13538, sta. 10, H = 43.8 mm.

9-12. Flabellum flexuosum n. sp.: 9, USNM 47265, Eltanin sta. 418, H = 36.0 mm; 10-12, holotype, USNM 47170, Eltanin sta. 1536, GCD = 20.8 mm, H = 28.7 mm.

Discovery sta. 34, 39), pl. 2, figs. 5, 6.
Flabellum transversale: Thomson and Rennet, 1931, p. 41.
Flabellum harmeri; Gardiner, 1939, pp. 326, 327.
Flabellum curvatum; Gardiner, 1939, pp. 327, 328 (part: Discovery sta. 182).
Frabellum marmeri; Niino, 1958, p. 257, pl. 2, fig. 1 (misspellings).
Flabellum sp. Squires, 1961, pp. 29, 38.
Flabellum impensum Squires, 1962b, pp. 14, 17-19, pl. 2, figs. 4-7, text fig. 3; 1969, p. 18 (part), pl. 6, map 3.--Podoff, 1976, pp. 31-38, pl. 2, figs. 9, 10.--Sorauf and Podoff, 1977, p. 4 (part: pl. 1, figs. 1, 2, not pl. 3, figs. 1, 2)--Cairns, 1979, p. 206.
Flabellum distinctum; Eguchi, 1965, pp.. 10, 11, text fig. 3.

Description. Corallum variable in shape, ranging from flabellate to almost conical (ceratoid to trochoid). Largest flabellate corallum 128 x 45 mm in CD and 80.2 mm tall (Plate 12, fig. 4). Ceratoid to trochoid coralla usually smaller, with GCD/LCD ratio as low as 1.1. Angle of lateral edges of flabellate coralla often about 40°-50° for first 10 mm, then lateral edges flaring outward up to 150°, maintaining constant inclination of lateral faces of 30°-40°. In ceratoid to trochoid coralla, angle of lateral edges remaining constant with growth. All intergrades occurring between extreme flabellate and ceratoid shapes. Round pedicel never reinforced by external stereome and 3.5-6.0 mm in diameter. Small coralla usually attached to pebbles or gastropod shells; when over 30 mm tall, coralla usually becoming free of original attachment and pedicel often eroding to a point. Theca thin and usually worn. Often a thin, incised costal line corresponding to each septum and closely spaced chevronate growth lines circling theca. Principal costae on lateral edges sometimes slightly carinate. Calicular profiles of flabellate coralla strongly arched; those of trochoid coralla more even.

Septa hexamerally arranged in smaller coralla up to 96 septa stage. Additional septa irregularly added in groups of three, up to at least 300 septa (S_7). Full fourth cycle achieved at GCD of about 11 mm, fifth cycle by about 19 mm, and sixth cycle by about 55 mm. Flabellate coralla usually having more septa than more slender coralla because of their increased calicular perimeter. In corallum with six full cycles of septa, relative septal sizes may be either $S_1 = S_2 = S_3 = S_4 \geq S_5 \geq S_6$ or $S_1 = S_2 = S_3 \geq S_4 \geq S_5 \geq S_6$. Septa not exsert and very thin. Larger septa usually sloping concavely away from calicular edge, producing shallow notch, this notch usually finely dentate. Toward center of calice, septum forming shoulder before dropping vertically into fossa. Inner edges of larger septa slightly sinuous and thickened; deep within fossa, lower inner edges fused into rudimentary columella. Septal granules fine and pointed, often arranged in widely spaced lines parallel to septal edge.

Fossa very deep, elongate in flabellate coralla. Small amounts of internal stereome sometimes present in base.

Remarks. An analysis of the living orientation of F. impensum similar to that of Squires [1964a] for other flabellids was made. The presence of filter feeders (usually Bryozoa) on both sides of a specimen was interpreted as an indication of an upright position, filter feeders on one side and borers on the other as a prone position, and absence of organisms from one or both sides as inconclusive. In an examination of 96 specimens which were alive when they were collected, 71 were found to be inconclusive, 24 indicated an erect posture, and 1 indicated a prone posture. The implication is that after F. impensum becomes free of its original attachment, it maintains itself in an upright position, perhaps by submergence of the pedicel in the substrate. Consistent absence of filter feeders from near the base and of erosion of the base supports this theory.

Two gastropods, believed to be of the families Volutidae and Muricidae (J. Houbrick, personal communication, 1979), have deposited large egg cases near the calicular edges of living specimens (Plate 12, figs. 4 and 5). Bryozoans and serpulids also commonly encrust the theca.

Pax [1910] describes and illustrates some histology of a tentacle, including nematocysts.

Discussion. The extreme forms of F. impensum might easily be mistaken for separate species: one a large flabellate corallum with almost seven cycles of septa, the other a ceratoid to trochoid corallum with less than six cycles of septa. Furthermore, each form usually occurs homogeneously when it is collected. However, several suites show a continuous gradation of corallum shape. No other morphological differences are apparent. Even the type-specimens show this corallum variation: the holotype is flabellate, and the illustrated paratype is ceratoid. Finally, there are no geographic or bathymetric differences, except that the few records from the Weddell Sea and Eastern Hemisphere are mostly slender forms. The variation in corallum shape is believed to be a response to a localized environmental factor which could vary over a short distance, such as substrate type or current pattern.

Early records of F. impensum were usually assigned to more northern temperate species. F. inconstans Marenzeller, 1904, reported by Pax [1910], is a South African species with a basal fracture. F. thouarsii Milne Edwards and Haime, 1848, is a distinct species known from relatively shallow water only off eastern South America. F. harmeri Gardiner, 1929, has been synonymized with the New Zealand F. rubrum by Squires [1963b], and F. transversale Moseley, 1881, reported by Thomson and Rennet [1931], is a southern Australian species. Gardiner's [1939] single Antarctic record of F. curvatum is F. impensum.

Material. Eltanin sta. 272 (6), USNM 45636; sta. 410 (4), USNM 45634; sta. 426 (3), USNM 45635; sta. 428 (5), USNM 45632; sta. 437 (5), USNM 45637; sta. 444 (1), USNM 47330; sta. 499 (1), USNM 47347; sta. 992 (7), USNM 47335; sta. 993 (2), USNM 47323; sta. 997 (10), USNM 47341; sta. 1002 (1), USNM 47327; sta. 1079 (7), USNM 45640; sta. 1083 (9), USNM 47331; sta. 1084 (3), USNM 47368; sta. 1089 (1), USNM 47542; sta. 1870 (8), USNM 47346; sta. 1871 (33), USNM 47361, and (1), MCZ; sta. 1878 (12), USNM 47356; sta. 1880 (6), USNM 47349; sta. 1885 (2), USNM 47344; sta. 1898 (1), USNM 47367; sta. 1916 (1), USNM 47362; sta. 1922 (1), USNM 47545; sta. 1930 (2), USNM 47339; sta. 1933 (3), USNM 47355; sta. 1996 (1), USNM 47343; sta. 2005 (2), USNM 47371; sta. 2006 (1), MCZ; sta.

2016 (4), USNM 47370, and (3), MCZ; sta. 2018 (3), USNM 47375, and (1), MCZ; sta. 2021 (1), USNM 47369; sta. 2025 (2), USNM 47357; sta. 2031 (1), USNM 47372; sta. 2045 (3), USNM 47354; sta. 2063 (4), USNM 47353; sta. 2065 (2), USNM 47352; sta. 2068 (1), USNM 47351; sta. 2070 (1), USNM 47540; sta. 2085 (2), USNM 47348; sta. 2088 (4), USNM 47350; sta. 2097 (1), USNM 47342; sta. 2099 (6), USNM 47358; sta. 2115 (1), USNM 47365; sta. 2117 (9), USNM 47374; sta. 2124 (9), USNM 47376; sta. 2143 (9), USNM 47373; sta. 5761 (1), USNM 47366; sta. 5765 (4), USNM 47326. Islas Orcadas sta. 575-53 (11), USNM 47383; sta. 575-65 (9), USNM 47336; sta. 575-66 (4), USNM 47379; sta. 575-67 (1), USNM 47380; sta. 575-70 (2), USNM 47384; sta. 876-107 (3), USNM 47334; sta. 876-108 (6), USNM 47378; sta. 876-110 (2), USNM 47332; sta. 876-113 (2), USNM 47381; sta. 876-114 (3), USNM 47324; sta. 876-118 (6), USNM 47337; sta. 876-124 (1), USNM 47329; sta. 876-126 (4), USNM 47340; sta. 876-127 (1), USNM 47382. Hero sta. 691-20 (15), USNM 47333; sta. 702-465 (1), USNM 47543; sta. 721-1084 (1), USNM 47325; sta. 721-1102 (1), USNM 47390; sta. 721-1110 (20), USNM 47363; sta. 731-1842 (1), USNM 47328. Edisto sta. 16 (1), USNM 47391; sta. 28 (2), USNM 45641. Atka sta. 23 (2), USNM 47345. EW sta. 9 (86), USNM 45629; sta. 16 (1), USNM 47385; sta. 23 (73), USNM 45627; sta. 28 (3), USNM 45644; sta. 32 (1), USNM 47387; sta. 35 (1), USNM 47386; sta. 36 (2), USNM 45667; sta. 37 (8), USNM 45666; sta. 38 (13), USNM 45630; sta. 39 (4), USNM 45628. Westwind sta. 4 (1), USNM 47389. Vema sta. 18-32 (2), USNM 47338. NZOI sta. A-537 (1), USNM 47364; sta. A-625 (20), USNM 47392. Specimen identified as F. transversale from station 10 [Thomson and Rennet, 1931], Australian Museum G 13538; specimen identified as F. harmeri [Gardiner, 1939], BM 1939.7.20.128; specimens (2) identified as F. curvatum from station 182 [Gardiner, 1939], BM 1939.7.20.271-272; specimens identified as F. thouarsii from Wells [1958] from stations 34 (2) and 39 (2), South Australian Museum H 63, H 65.

Types. The holotype and 61 paratypes of F. impensum are deposited at the New Zealand Oceanographic Institute. Type-locality: 73°20'S, 174°00'E (Ross Sea); 369-384 m.

Distribution. Circumpolar continental Antarctic, including off South Shetland Islands, South Orkney Islands, and South Sandwich Islands, and one disjunct record off the Antipodes Islands (Map 8). Depth range: 46-2260 m; however, temperature range probably very slight. Most records between 100 and 1000 m; one of deepest records (2010 m) at northernmost range (Eltanin sta. 2143, off Antipodes Islands).

26. Flabellum flexuosum n. sp.
Plate 12, figs. 9-12

Desmophyllum sp. Marenzeller, 1903, p. 1.
Desmophyllum antarcticum; Gravier, 1914b, p. 122 (part: pl. 1, fig. 4).
Gardineria lilliei; Gardiner, 1939, pp. 328, 329 (part: Discovery sta. 140, 160 (part), 181, 190).
Flabellum antarcticum; Wells, 1958, p. 269, pl. 2, figs. 11-15.--Squires, 1962b, pp. 13, 14, 19, 20; 1969, p. 18, pl. 6, map 3.--Bullivant, 1967, p. 65.--Zibrowius, 1974b, p. 18.--Not F. antarcticum; Keller, 1974, p. 203 (is F. curvatum Moseley, 1881).

Flabellum thouarsii; Wells, 1958, p. 268 (part: Discovery sta. 41, 93), pl. 2, figs. 7-10.
Flabellum ongulense Eguchi, 1965, pp. 11, 12, pl. 2, figs. 2a-2d.

Description. Corallum ceratoid, tall; straight, bent, curved, or scolecoid. Pedicel 2.7-4.5 mm in diameter, expanding slightly (up to 5.5 mm) at attachment to substrate. Coralla usually remaining attached. Holotype 20.8 x 17.7 mm in CD and 28.7 mm tall: PD at break 3.0 mm. Largest specimen (Eltanin station 1933, USNM 47172) 24.0 x 20.7 mm in CD, 4.0 mm in PD, and 67 mm tall. Theca very thin and porcelaneous, usually without encrusting organisms; however, bryozoans sometimes colonizing theca of living specimens. Calice elliptical, not compressed; ratio of GCD/LCD about 1.25.

Septa hexamerally arranged in five cycles; however, tall, slender coralla and younger specimens often with less septa. S_1 and S_2 equal in size and slightly exsert (because the septa are so delicate, their upper septal edges are invariably broken off when they are collected). Remaining septal cycles progressively smaller, S_3 sometimes 3-4 times larger than S_4; S_5 rudimentary. Inner edges of all septa sinuous, corresponding to shallow, transverse undulations on septa, producing wrinkled or corrugated appearance. Fine, pointed septal granules, up to 2 times septal thickness in height, aligned on crests of septal undulations. Lower inner edges of S_1 and S_2 usually fused, forming rudimentary columella.

Discussion. Both Wells [1958] and Squires [1962b, 1969] identified this species as Flabellum antarcticum (Gravier, 1914). Gravier's [1914a] species, although it is very similar to F. flexuosum, is a rarely collected Javania, which always has a thick stereome-reinforced pedicel; F. flexuosum always has a typical nonreinforced Flabellum-type pedicel. However, Gravier's [1914b, pl. 1, fig. 4] third specimen from dredge VIII, which he doubtfully assigned to Desmophyllum antarcticum, is probably F. flexuosum.

F. flexuosum has basically the same distribution as F. impensum, the only other circum-Antarctic species of Flabellum. It can be distinguished by its more slender, often bent corallum, exsert septa, and more prominent septal granulation.

The single specimen from the Weddell Sea (Edisto station 20) and Wells's [1958] specimens from off eastern Antarctica differ from typical F. flexuosum in that they are large specimens and yet have only four cycles of septa. It would be helpful to have more specimens from the eastern Antarctic for comparison.

Flabellum ongulense Eguchi, 1965, may be the same species and would therefore have nomenclatural priority. However, none of Eguchi's Antarctic specimens are available for study. F. gracile (Studer, 1978), known only from off New Zealand (95-196 m), is also very similar to F. flexuosum. Thomson and Rennet's [1931] Caryophyllia vermiformis seems to be F. gracile.

Etymology. The specific name flexuosum (Latin: full of bends, crooked) refers to the bent and often scolecoid shape of the corallum.

Material. Eltanin sta. 418 (1), USNM 47265; sta. 671 (42), USNM 47275; sta. 684 (5), USNM 45656; sta. 993 (2), USNM 53427; sta. 1535 (7), USNM 47270; sta. 1870 (3), USNM 47259; sta. 1878 (1), USNM 47261; sta. 1995 (22), USNM 47276; sta.

1996 (26), USNM 47282; sta. 1997 (10), USNM 47279; sta. 2092 (1), USNM 47268; sta. 2097 (1), USNM 47258; sta. 2119 (1), USNM 47257; sta. 2120 (1), USNM 47256. Islas Orcadas sta. 575-8 (1), USNM 47277; sta. 575-12 (4), USNM 47283; sta. 575-13 (1), USNM 47273; sta. 575-17 (3), USNM 47267; sta. 575-30 (1), USNM 47262; sta. 575-52 (1), USNM 47273; sta. 575-90 (1), USNM 47281; sta. 575-91 (6), USNM 47271. Hero sta. 691-20 (1), USNM 47280; sta. 731-1812 (1), USNM 47266. Edisto sta. 20 (2), USNM 47272. Atka sta. 23 (12), USNM 47269. Burton Island sta. 3 (3), USNM 47264. EW sta. 6 (2), USNM 47255; sta. 28 (3), USNM 47274; sta. 35 (1), USNM 47263. At 66°40'S, 139°51'E, 220-240 m (1), SME. Specimen identified as Desmophyllum sp. by Marenzeller [1903], Brussels Museum; specimen identified as Caryophyllia vermiformis by Thomson and Rennet [1931], Australian Museum G 13535; specimens (5) identified as Gardineria lilliei by Gardiner [1939], BM 1939.7.20.238-240, and (1), MCZ; specimens identified as F. thouarsii from stations 41 (5) and 93 (3) [Wells, 1958], South Australian Museum H 66, H 69. Types.

Types. The holotype, collected at Eltanin station 1536, is deposited at the United States National Museum (47170). Nine paratypes from Eltanin station 1536 (number 47171), 1 from Eltanin station 1933 (number 47172), and 21 from Islas Orcadas station 575-93 (number 47173) are deposited at the United States National Museum. Type-locality: 54°29'S, 39°22'W (west of South Georgia); 659-686 m.

Distribution. Off Antarctic Peninsula; off South Shetland Islands; off South Orkney Islands; off South Sandwich Islands; off South Georgia and Shag Rocks; Weddell Sea; off Enderby Land; Ross Sea; Bellingshausen Sea (probably circumpolar) (Map 10). Depth range: 101-659 m.

27. Flabellum gardineri n. sp.
Plate 13, figs. 1-3

Gardineria lilliei; Gardiner, 1939, pp. 328, 329 (part: 40 specimens from Discovery sta. 160).

Description. Corallum ceratoid, straight, elongate. Pedicel diameter about 2.5 mm, expanding slightly at attachment to substrate. Holotype 8.9 x 8.3 mm in CD and 22.1 mm tall. Tallest specimen 30.5 mm. Theca dull white with thin, incised vertical striae, one corresponding to each septum. Calice round or elliptical.

Septa hexamerally arranged in four cycles. S_1 and S_2 equal in size and 3-4 times larger than S_3 and S_4, these about equal in size. Septa not exsert; larger septa bearing nondentate, shallow notch near calicular edge. Inner septal edges of S_1 and S_2 straight and entire, fusing into a solid columella about one third to one half of distance to base. S_3 and S_4 very low in relief with irregular inner edges. Septal granules sparse, small, and pointed.

Discussion. Gardiner [1939] reported 40 specimens of Gardineria lilliei from four stations and implied that 4 environmentally controlled forms were present. Forty-nine specimens, deposited at the British Museum and the Museum of Comparative Zoology, have been examined from these stations, all bearing Gardiner's identification of G. lilliei; however, the 4 implied growth forms are in fact separate species belonging to two genera,

neither of which is Gardineria: Flabellum flexuosum (Discovery stations 140, 160 (part), 181, and 190); Flabellum gardineri (Discovery station 160 (part)); Flabellum sp. (Discovery station 160 (part)); and Caryophyllia eltaninae (Discovery station 160 (part)). The 2 specimens from Discovery station 160 referred to above as Flabellum sp. differ from Flabellum gardineri in having a wider pedicel, a more open calice, and a fifth cycle of septa; otherwise they are very similar and may represent a growth form. Flabellum sp. is also represented by Gardiner's [1929a] record of G. antarctica from Discovery station 152. More specimens are required before this problem can be solved.

Flabellum gardineri most closely resembles Flabellum flexuosum but can be distinguished by its fewer septa, more massive columella, straight inner septal edges, noncorrugated septa, and smaller size.

Etymology. This species is named in honor of J. S. Gardiner, who contributed greatly to our knowledge of Scleractinia, including corals of the Subantarctic region.

Material. Discovery sta. 160 (4), MCZ-3574. Types.

Types. The holotype, collected at Discovery station 160, is deposited at the British Museum (1939.7.20.305). Thirty-three paratypes from Discovery station 160 (1939.7.20.288-304, 306-314) are deposited at the British Museum. Two specimens from this lot have been permanently deposited at the United States National Museum (48300). Type-locality: 53°43'40"S, 40°57'00"W (off Shag Rocks); 177 m.

Distribution. Known only from type-locality (Map 9).

28. Flabellum knoxi Ralph and Squires, 1962
Plate 13, figs. 4-7

Flabellum knoxi Ralph and Squires, 1962, pp. 14, 15, pl. 7, figs. 1, 2.--Squires, 1964a, pp. 11, 12, 19, 20, pl. 1, figs. 4-6, pl. 2, fig. 7, pl. 3, figs. 3-5, pl. 4, figs. 1-4; 1969, p. 18, pl. 6, map 4.--Squires and Keyes, 1967, pp. 26, 27, pl. 5, figs. 1, 2.--Zibrowius, 1974b, p. 18.

Description. Corallum flabellate, compressed; angle of lateral edges typically 135°-180°, inclination of lateral faces 30°-35°. Lateral edges usually rounded, not carinate. Base of pedicel small, 2.5-3.0 mm in diameter; height about 5-10 mm. One of largest specimens (holotype) 112 x 55 mm in CD and 65 mm tall. Theca very thin and fragile, bearing thin, incised striae, one corresponding to each septum. Closely spaced, transverse growth lines form chevrons, peaking at each stria. Theca uniformly reddish-brown, entirely white, or bearing reddish-brown stripes corresponding to each septum, Darker, broader stripes corresponding to major septa. Calice entire, not lacerate, and strongly arched.

Up to 348 thin, fragile septa per calice, arranged in three size groups. Largest septa (primaries) extending to columella and having very sinuous lower inner edges. Between each primary a secondary, usually smaller (three fourths of a size of primary) but in larger specimens almost reaching columella. A much smaller tertiary septum, occurring between each primary and secondary, rarely extending more than halfway to pedicel. In

Plate 13. Flabellum

1-3. Flabellum gardineri n. sp.: holotype, BM 1939.7.20.305, Discovery sta. 160, H = 22.1 mm, GCD = 8.9 mm.

4-7. Flabellum knoxi Ralph and Squires: 4, USNM 53378, NZOI sta. D-177, GCD = 44.0 mm; 5, USNM 47492, NZOI sta. D-175, H = 22.7 mm, stripes on theca; 6, USNM 53380, NZOI sta. D-179, H = 42.7 mm, coated with ammonium chloride; 7, USNM 47492, NZOI sta. D-175, x1.7, corallum broken, revealing columella.

8-11. Flabellum apertum Moseley: 8, 9, lectotype, BM 1880.11.25.74, Challenger sta. 145, GCD = 32.2 mm; 10, 11, USNM 47444, Eltanin sta. 1412, GCD = 57.3 mm.

larger specimens, rudimentary quaternaries present near calicular edge. Upper septal margins invariably broken, but septa do not appear to be exsert. Very small, pointed granules arranged in widely spaced rows parallel to septal margins. Columella long and slender (1.5-2.2 mm wide) but very sturdy, often remaining intact after surrounding septa have been broken away (Plate 13, fig. 7). Columella composed of loose fusion of convoluted lower inner edges of primary septa.

Remarks. The living coral appears to remain in the upright position, as is concluded from the presence of attached filter feeders on both lateral faces. Squires [1964a] suggested that it maintains this orientation by sinking its pedicel into soft mud in areas of low-velocity current. Coralla are rarely attached, but if they are, to sand or pebbles, which are usually incorporated into the pedicel. According to Squires [1974a], specimens that are accidentally knocked to the prone position will produce recurved (angle of lateral edges up to 250°) and reflexed coralla in an effort to right the upper half of the polyp relative to the substrate.

The variation in thecal striping and color of F. knoxi is similar to that found in F. pavoninum atlanticum Cairns, 1979. In general, the intensity of striping correlates with age [Squires and Keyes, 1967], the younger specimens having the more pronounced stripes.

Discussion. F. knoxi is extremely similar and may be identical to F. magnificum Marenzeller, 1904, from off Sumatra. The type of the latter species was not examined.

Material. Eltanin sta. 1398 (2), USNM 47496; sta. 1989 (7), USNM 47493. NZOI sta. A-898 (5), USNM 47495; sta. D-6 (2), USNM 47494; sta. D-175 (4), USNM 47492; sta. D-176 (15), USNM 53379; sta. D-177 (6), USNM 53378; sta. D-179 (2), USNM 53380; sta. D-207 (8), USNM 53376.

Types. The holotype is deposited at the Canterbury Museum, Christchurch, New Zealand. Type-locality: Chatham Rise; 402-512 m.

Distribution. Chatham Rise; Campbell Plateau; off Macquarie Island (Map 9). Depth range: 201-914 m.

29. Flabellum apertum Moseley, 1876
Plate 13, figs. 8-11; Plate 14, figs. 1-4

Flabellum apertum Moseley, 1876, p. 556 (part: off Prince Edward Islands); 1881, pp. 167, 168 (part: Challenger sta. 145), pl. 6, figs. 7a-7c.--Not F. apertum; Marion, 1906, pp. 120, 121, pl. 11, figs. 9, 9a (is F. angulare Moseley, 1876).--Wells, 1958, p. 262.--Not F. apertum Squires, 1958, p. 68.--Squires and Keyes, 1967, p. 26, pl. 4, figs. 4, 5.--Squires 1969, pp. 16, 18, pl. 6, map 4.--Zibrowius, 1980, p. 154.

Flabellum patagonichum Moseley, 1881, pp. 166, 167, pl. 15, figs. 1-7.--Fowler, 1885, pp. 585-590, figs. 1-12.--Wells, 1958, p. 262. --Squires, 1961, p. 30; 1969, pp. 17, 18, pl. 6, map 4.--Squires and Keyes, 1967, p. 27.--Cairns, 1979, p. 206.

Description. Corallum campanulate and distinctly compressed. Pedicel short and cylindrical (2.0-2.5 mm in diameter); originally attached to small object, becoming free early in ontogeny. Largest specimen examined (USNM 47444) 57.2 x 39.3 mm in CD and 37.2 mm tall. The two principal costae ridged and continuous from pedicel to calice. They diverge from pedicel at an apical angle between 130° and 170° until a GCD of about 30 mm, whereupon epitheca turns upward to continue almost vertical growth. At point of inflection, the four lateral C_1 usually well developed, sometimes forming spurs, and may continue as ridges to calice. C_2 sometimes ridged from point of inflection to calice, but much less than C_1. Epitheca porcelaneous, with chevron-shaped growth lines extending between each septum. Calicular profile scalloped, a large apex corresponding to each S_1 and S_2 and a smaller peak corresponding to every S_3.

Septa hexamerally arranged in four cycles, with rudiments of fifth cycle only in larger specimens. S_5 first appearing in end half systems, only rarely in lateral half systems; largest specimen with 68 septa. S_1 and S_2 equal in size and slightly exsert. Their inner edges thickened and fusing in center of fossa, forming elongate, solid or trabecular columella. S_3 and S_4 progressively smaller and do not reach columella. S_5, if present, rudimentary. All septa having straight inner edges and bearing numerous, small, pointed granules. Stereome infilling sometimes at bottom of fossa.

F. apertum forma patagonichum differing from typical form primarily in its smaller size (maximum size reported, 28 x 21 mm in CD) and more slender shape. Ridged principal costae diverging from pedicel at an apical angle between 90° and 110° and turning upward at GCD of about 15 mm, producing smaller, more slender corallum. Other four C_1 not developed. Theca porcelaneous only in region 5-10 mm from calice. Remainder of corallum usually worn and white but may also be uniformly reddish-brown or white with diffuse reddish-brown stripes corresponding to S_1 and S_2. S_4 often missing from half systems and stereome infilling more common.

Discussion. F. patagonichum is treated as a forma of F. apertum because both patagonichum and typical apertum as well as a continuous series of morphological intermediates were present in two lots. The series of 246 specimens from Eltanin station 283 was particularly helpful in tracing the morphological variation possible in one population.

Four other closely related species have been linked to F. apertum: F. angulare Moseley, 1876; F. conuis Moseley, 1881; F. japonicum Moseley, 1881; and F. raukawaensis Squires and Keyes, 1967. Gardiner [1929b] synonymized F. angulare, F. apertum, and F. conuis as F. japonicum, the incorrect senior synonym. He stated that F. patagonichum might prove to be a form of F. japonicum also. Keller [1974] synonymized F. japonicum and F. raukawaensis as F. apertum, which she reported from off South Africa and western India. Zibrowius [1980] distinguished F. angulare from F. apertum and discussed the nominal species.

I have examined the type-specimens of all of the above species except F. raukawaensis, of which I have seen a specimen from very near the type-locality (Eltanin station 1403), and have made the following observations. F. japonicum can be distinguished from F. apertum by its possession of a full fifth cycle of septa at a calicular diameter at which F. apertum has only few S_5. Yabe and

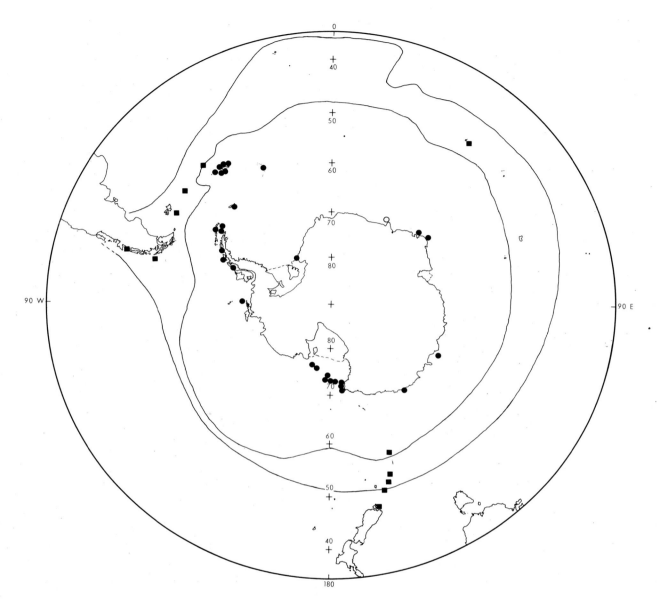

Map 10. Distribution of _Flabellum flexuosum_ (solid circles), _Flabellum ongulense_ (open circle), and _Flabellum apertum_ (solid squares).

Eguchi [1941a] have used this as a key character to differentiate the two species. _F. japonicum_ has been widely reported from the northern Indian Ocean, off Indonesia, Japan, Philippine Islands, and Tasmania. _F. angulare_ is known only from the north Atlantic (1647-3186 m) and can be distinguished by its less compressed corallum and S_3 that extend to the columella. _F. conuis_, known only from off the Admiralty Islands, is very similar to _F. apertum_ forma _patagonichum_ in shape but differs in having a deeper fossa, a more delicate corallum, slightly corrugated septa, and S_3 that reach the columella. It is probably a separate species, but obviously more specimens are needed before this can be determined with certainty. _F. raukawaensis_, known only from four specimens from off North Island, New Zealand, is very similar to typical _F. apertum_ but differs in that it has more S_5, a larger columella, and S_3 that almost reach the columella.

Other records of _F. apertum_ that I have not verified include von Marenzeller [1904a], off Tanzania, East Africa, 863 m; Yabe and Eguchi [1941b], off Japan, 307 m; Keller [1974], off South Africa, 1400 m, and off western India, depth unknown; Keller [1975], Caribbean, depth unknown.

Material. Forma apertum: Eltanin sta. 338 (2), USNM 45675; sta. 558 (1), USNM 47442; sta. 1412 (4), USNM 47444; sta. 1414 (2), USNM 47439; sta. 1422 (5), USNM 47441. Islas Orcadas sta. 575-6 (12), USNM 47445. Edisto sta. 7 (1), USNM 4736. WH sta. 64/68 (3), ZIZM. NZOI sta. D-166 (2), USNM 47437. Mixtures of formae apertum and patagonichum: Eltanin sta. 216 (28), USNM 45674; sta. 21-283 (246), USNM 47443. Specimen of Squires and Keyes [1967], USNM 47438. Syntypes of _F. apertum_ and _F. patagonichum_.

Types. The eight syntypes of _F. apertum_, two from Challenger station 3 (number 1880.11.25.73) and six from Challenger station 145 (number

1880.11.25.74), are deposited at the British Museum. Because these specimens represent a mixed lot [Zibrowius, 1980], a lectotype is chosen from Challenger station 145 (Plate 13, figs. 8, 9). The remaining specimens are considered paralectotypes, those from Challenger station 3 being F. angulare [see Zibrowius, 1980]. Type-locality: 46°40'S, 37°50'E (off Prince Edward Island); 567 m. The syntypes of F. patagonichum are also deposited at the British Museum. Type-locality: 47°48'30"S, 74°47'W (off Isla Pengüin, Chile); 220 m.

Distribution. Circum-Subantarctic, including off Prince Edward Islands; Hjort Seamount; Macquarie Ridge; off Chile; off Falkland Islands; Falkland Plateau; off southern Brazil; off southern New Zealand (Map 10). Depth range: 220–1500 m.

30. Flabellum truncum n. sp.
Plate 14, figs. 5–8

Description. Corallum ceratoid to trochoid, compressed. Angle of lateral edges 45°–70°; inclination of lateral faces 22°–38°; ratio of GCD/LCD 1.4–2.2. Corallum (anthocyathus) proximally truncated, resulting from transverse fission from a presumably attached base (the anthocaulus). Scar of attachment from 9 to 14 mm long and from 5 to 9 mm wide; usually worn. Largest specimen 38 x 23 mm in CD; tallest specimen 39.4 mm in height. Theca thin, sometimes bearing low, rounded, longitudinal ridges, one corresponding to each S_{1-3}. Lateral edges rounded, never carinate or spinose. Calicular margin entire; profile of margin arched.

Septa hexamerally arranged in five cycles, rarely with additional S_6. S_1, S_2, and S_3 equal in size and extending to columella. S_4 about half of size of S_1; S_5 about one fourth of size of S_4. Sometimes lower inner edges of S_4 bending toward and fusing with S_3. Septa not exsert and bearing large, pointed or blunt granules measuring as high as septal thickness. Lower inner edges of larger septa thickened and slightly sinuous; sinuosity corresponding to shallow undulations of currugated septa. Columella variable in structure, but usually slender, elongate, loose fusion of inner edges of S_{1-3}; may sometimes be a slightly wider, flat, solid mass or a very wide (up to 25% of LCD), loose, spongy structure.

Remarks. Zibrowius [1974b] suggested that the truncated flabellids may reproduce asexually by transverse division ('strobilation') with one basal part (the anthocaulus stage [see Wells, 1966, p. 226]), producing more than one corallum (the anthocyathus stage). Unfortunately, no attached specimens or specimens in the process of dividing were found among the lots of F. truncum. When the attached and free stages of the truncated flabellids are known, some species will probably be synonymized.

Discussion. F. truncum belongs to the flabella truncata section of the genus Flabellum [see Milne Edwards and Haime, 1848, p. 259], which is equivalent to Zibrowius's [1974b] 'second group.' Zibrowius listed 19 species from this group, which are all characterized by a transverse division, but none of these are known from the eastern Pacific or the Subantarctic. Many are distinguished by prominent costal spines or crests. Out of this group, F. truncum is most similar to F.

inconstans Marenzeller, 1904, known from off South Africa at 100 m. F. truncum is distinguished by its smaller size, lesser number of septa, and deeper bathymetric range.

Etymology. The specific name truncum (Latin: piece cut off, tip) refers to the detached distal anthocyathus stage of this species.

Material. Eltanin sta. 21-282 (1), USNM 47526; sta. 338 (1), USNM 47527. Islas Orcadas sta. 575-5 (1), USNM 47528. Anton Bruun sta. 11-88 (2), USNM 47529. Types.

Types. The holotype, collected at Eltanin station 21-283, is deposited at the United States National Museum (47174). Six paratypes from Eltanin station 21-283 (number 47175) and 19 from Anton Bruun station 18-714 (number 47176) are deposited at the United States National Museum. One paratype from Anton Bruun station 18-714 is deposited at the British Museum (1979.11.4.1). Type-locality: 53°13'S, 75°41'W (off Isla Desolación, Chile); 1500–1666 m.

Distribution. Off western coast of South America from off Peru to off southern Chile; south of Falkland Islands; Falkland Plateau (Map 11). Depth range: 595–1896 m.

Genus Javania Duncan, 1876

Diagnosis. Solitary, ceratoid to trochoid, fixed. Wall epithecal. Base reinforced by layers of stereome. No pali. Calicular edge jagged. Columella rudimentary. Type-species: Javania insignis Duncan, 1876, by monotypy.

31. Javania cailleti (Duchassaing and Michelotti, 1864)
Plate 14, figs. 9–12

Desmophyllum cailleti Duchassaing and Michelotti, 1864, p. 66, pl. 8, fig. 11.
Desmophyllum eburneum Moseley, 1881, p. 162, pl. 6, figs. 1a, 1b.
Desmophyllum nobile Verrill, 1885, pp. 150, 151.
Desmophyllum vitreum Alcock, 1898, p. 20, pl. 2, figs. 2a, 2b.
Flabellum sp. Marenzeller, 1904b, pl. 81.
Desmophyllum galapagense Vaughan, 1906b, p. 63, pl. 1, figs. 1a, 1b.
Javania eburnea; Zibrowius, 1974b, pp. 12, 13, pl. 3, figs. 13–17; 1980, pp. 157–159, pl. 82, figs. A–L.
Javania cf. eburnea; Zibrowius, 1974b, pp. 13–16, pl. 4, figs. 22–29, pl. 5, figs. 31–34.
Javania vitrea; Zibrowius, 1974b, pp. 16, 17, pl. 5, figs. 18–21.
Javania cailleti; Cairns, 1979, pp. 153–156, pl. 28, figs. 8–12, pl. 30, figs. 1, 4.

Description. J. cailleti has been fully described and illustrated elsewhere [Zibrowius, 1974b, 1980; Cairns, 1979]; only a brief diagnosis is given here. Corallum ceratoid, often flared distally. Pedicel thick, reinforced by concentric layers of stereome up to PD of one fourth to one half of CD. Typical specimen 18 x 14 mm in CD and 35 mm tall. Theca usually smooth and porcelaneous but may be ridged with costae near calice. Septa usually hexamerally arranged in four cycles: S_1 = S_2 ≥ S_3 ≥ S_4 in size. S_1 and S_2 highly exsert; S_4 rudimentary. Inner septal edges straight. Septal granules low, rounded.

Plate 14. _Flabellum_ and _Javania_

1-4. _Flabellum apertum_ Moseley: 1, USNM 47443, _Eltanin_ sta. 283, H = 24.0 mm; 2, specimen from same lot, GCD = 26.8 mm; 3, 4, syntype of F. patagonichum Moseley, BM, _Challenger_ sta. 305, GCD about 28 mm.

5-8. _Flabellum truncum_ n. sp.: 5, 6, holotype, USNM 47174, _Eltanin_ sta. 283, H = 31.7 mm, GCD = 27.7 mm; 7, paratype, USNM 47175, _Eltanin_ sta. 283, GCD = 25.6 mm, large spongy columella; 8, USNM 47529, Anton Bruun sta. 11-88, x2.2, view of basal fracture.

9-12. _Javania cailleti_ (Duchassaing and Michelotti): 9, 10, USNM 19173, Albatross sta. 2785, H = 22.4 mm, GCD = 20.2 mm; 11, 12, USNM 47530, _Eltanin_ sta. 1592, H = 20.4 mm, GCD = 10.2 mm.

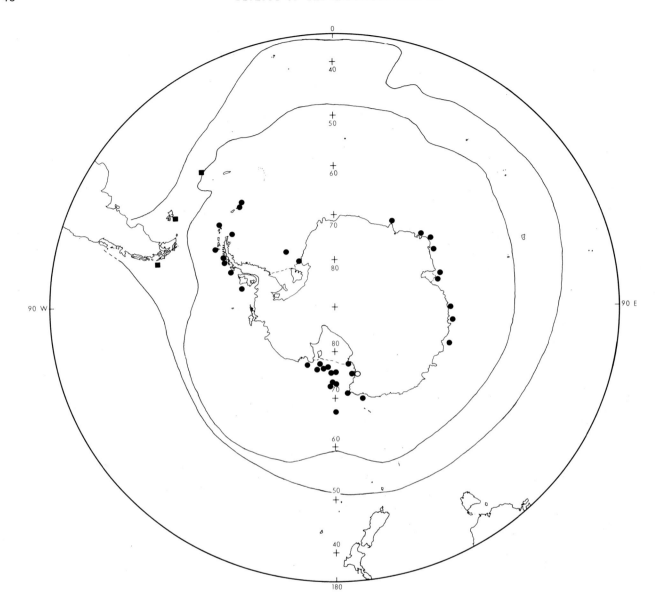

Map 11. Distribution of <u>Gardineria antarctica</u> (solid circles), fossil <u>Gardineria ant-</u>
<u>arctica</u> (open circle), and <u>Flabellum truncum</u> (solid squares).

Fossa deep and narrow; rudimentary columella formed by solid fusion of lower inner edges of S_1 and S_2.

Discussion. Of the two new specimens reported, the one from <u>Albatross</u> station 2785, about 10 km from the type-locality of <u>D</u>. <u>eburneum</u> Moseley, is identical to the syntypes of that species. The small specimen from <u>Eltanin</u> station 1592, however, has prominent costal ridges and highly exsert septa, similar to those of the holotype of <u>D</u>. <u>nobile</u> Verrill.

Material. <u>Eltanin</u> sta. 1592 (1), USNM 47530; <u>Albatross</u> sta. 2785 (1), USNM 19173. Specimens listed by Cairns [1979], USNM. Syntypes of <u>D</u>. <u>eburneum</u>; holotypes of <u>D</u>. <u>nobile</u> and <u>D</u>. <u>galapa-</u> <u>gense</u>.

Types. See Cairns [1979]. Type-locality: Lesser Antilles.

Distribution. Widespread in Northern Hemisphere in all oceans: off Galapagos Islands; off Japan; off India; off Morocco; northwest Mediterranean; Celtic Sea; off Madeira; off Azores; off Nova Scotia; off Georgia, United States, to off Suriname (including Gulf of Mexico and Caribbean). In Southern Hemisphere known from off Uruguay; off Burdwood Bank; off Chile (Map 12). Depth range: 86-2165 m.

32. <u>Javania antarctica</u> (Gravier, 1914) n. comb.
Plate 15, figs. 1-4

<u>Desmophyllum</u> <u>antarcticum</u> Gravier, 1914a, pp. 236-238; 1914b, pp. 122-125 (part: not dredge 8), pl. 1, figs. 1-3.
Not <u>Flabellum antarcticum</u>; Wells, 1958, p. 269 (is <u>F</u>. <u>flexuosum</u> n. sp.).--Not <u>F</u>. <u>antarcticum</u>; Squires, 1962b, pp. 13, 14, 19, 20 (is <u>F</u>. <u>flex-</u> <u>uosum</u> n. sp.) and not 1969, p. 18 (is <u>F</u>. <u>flexuo-</u> <u>sum</u> n. sp.).--Not <u>F</u>. <u>antarcticum</u>; Keller, 1974, p. 203 (is <u>F</u>. <u>curvatum</u>).

Plate 15. Javania and Gardineria

1-4.   Javania antarctica (Gravier): 1, MNHNP (no number), locality unknown, H =
        64.0 mm; 2, 3, USNM 47464, Hero sta. 731-1865, GCD = 24.7 mm; 4, USNM 53407,
        Edisto sta. 15, H = 71.7 mm.

5-11.   Gardineria antarctica Gardiner: 5, USNM 47249, Eltanin sta. 2119, H = 22.6
        mm; 6, specimen from same lot, CD = 27.4 mm, columella labyrinthiform; 7,
        USNM 47247, Eltanin sta. 5762, CD = 25.5 mm, coated with ammonium chloride,
        paliform lobes present; 8, USNM 47248, Eltanin sta. 2082, CD = 28.1 mm; 9,
        USNM 47205, Eltanin sta. 2117, CD = 26.1 mm, massive columella; 10, lecto-
        type of G. lilliei Gardiner, BM 1929.10.22.9, Terra Nova sta. 194, GCD =
        12.5 mm; 11, syntype of G. antarctica, BM Terra Nova sta. 349, CD = 19.0 mm.

? Desmophyllum delicatum; Niino, 1958, p. 257, pl. 2, fig. 3.
? Desmophyllum pseudoseptata Eguchi, 1965, p. 9, pl. 2, figs. 3a-3c.

Description.   Corallum ceratoid to trochoid, tall, straight to slightly curved.   Pedicel reinforced by numerous concentric layers of stereome, producing PD ranging from 5.7 to 12.5 mm.   Pedicel more expanded at base of attachment.   Largest specimen (Gravier's [1914b] illustrated syntype) 44 x 38 mm in CD, about 8.0 in PD, and 65 mm tall. Theca very thin and porcelaneous; encrusting organisms may settle on theca, these organisms periodically covered over with stereome.   Chevron-shaped growth lines peak at insertion lines of major septa.   Insertion lines sometimes slightly grooved, resembling costae.   Calice elliptical, ratio of GCD/LCD between 1.15 and 1.60.

Septa hexamerally arranged in five cycles. S_1 and S_2 equal in size and slightly exsert. S_3 slightly smaller but much larger than S_4; S_5 rudimentary.   Septa very thin and delicate; specimens rarely collected with intact upper septal edges.   Inner septal edges slightly sinuous, those of S_1 and S_2 thickened lower in fossa, there loosely fused into rudimentary collumella.   Septal granulation fine and pointed.   Fossa deep.

Discussion.   J. antarctica is very similar to F. flexuosum but can be distinguished by its thicker, reinforced pedicel and its later development of a full fifth cycle of septa.   Wells [1958] and Squires [1962b, 1969], perhaps not realizing the importance of the pedicel (Zibrowius resurrected Javania only as recently as 1974), lumped the two together.   Eguchi's [1965] Desmophyllum pseudoseptata is probably a Javania and may be J. antarctica; however, Eguchi's holotype could not be located.   Niino's [1958] D. delicatum is based on the same specimen.

Material.   Eltanin sta. 499 (1), USNM 47462; sta. 1054 (1), USNM 47455; sta. 1081 (1), USNM 47461; sta. 1089 (2), USNM 47458.   Islas Orcadas sta. 575-34 (1), USNM 47463; sta. 575-89 (1), USNM 47459.   Hero sta. 731-1865 (1), USNM 47464; sta. 731-1940 (1), USNM 47456; sta. 731-1947 (3), USNM 47460.   Edisto sta. 15 (3), USNM 53407.   EW sta. 4 (1), USNM 47457; one specimen without locality data (MNHNP).

Types.   One syntype from the Pourquoi-Pas? station 4 (illustrated by Gravier [1914b, pl. 1, figs. 1, 2]), is deposited at the Muséum National d'Histoire Naturelle, Paris.   The other syntype from the same station and the doubtfully assigned specimen from Pourquoi-Pas? station 8 (the latter probably F. flexuosum) could not be found at the Muséum National d'Histoire Naturelle; however, another large, typical specimen without locality data is present.   Type-locality: 64°50'S, 63°30'W (off Anvers Island, Palmer Archipelago); 53 m.

Distribution.   Off western Antarctic Peninsula; Scotia Ridge from South Shetland Islands to South Georgia; off Cape Norvegia, Queen Maud Land; ? off Riiser-Larsen Peninsula (Cape Cook), Prince Harald Coast (Map 12).   Depth range: 53-1280 m.

Genus Gardineria Vaughan, 1907

Diagnosis.   Solitary, turbinate to cylindrical, fixed.   Wall epithecal but thickened internally by stereome.   Septa not always arranged hexamerally. Paliform lobes opposite larger septa.   Columella well developed, papillose.   Type-species: Gardineria hawaiiensis Vaughan, 1907, by original designation.

33.   Gardineria antarctica Gardiner, 1929
Plate 15, figs. 5-11

Flabellum sp. Pax, 1910, p. 73, pl. 11, fig. 2.
Caryophyllia sp. ? Gravier, 1914b, pp. 130, 131, pl. 1, figs. 9, 10.
Gardineria antarctica Gardiner, 1929a, pp. 124, 125, 128-130, pl. 1, figs. 11-18; 1939 (part: not Discovery sta. 152), p. 328.--Wells, 1958, pp. 269, 270, pl. 2, figs. 16-18.--Squires, 1961, p. 20; 1962b, pp. 11, 13, 15, 20, 21, pl. 1, figs. 1-10; 1969, pp. 17, 18, pl. 6, map 2.--Speden, 1962, p. 756, figs. 11a-11c.--Zibrowius, 1974b, p. 24.--Podoff, 1976, pp. 45, 46, pl. 3, figs. 12, 13.
Gardineria lilliei Gardiner, 1929a, p. 125, pl. 1, figs. 3-10.--Wells, 1958, p. 262.
Caryophyllia inskipi; Thomson and Rennet, 1931, p. 40, pl. 10, fig. 6.
Ceratorochus (Convtorochus) parphis; Niino, 1958, p. 257, pl. 2, fig. 5 (misspellings).
Gardineria lillei; Squires, 1961, p. 20; 1962b, p. 13; 1969, pp. 17, 18, pl. 6, map 2.--Zibrowius, 1974b, p. 24.
Ceratotrochus ? sp. Eguchi, 1965, p. 8, pl. 2, figs. 1a, 1b.

Description.   Corallum usually straight, regular cone with round calice; trochoid to turbinate, sometimes ceratoid.   Pedicel diameter 2.8-7.7 mm, expanding up to 12 mm at base of attachment. Largest specimen examined 32.8 mm in CD and 32.2 mm tall.   Theca thick and not porcelaneous; usually smooth, but sometimes bearing low granulated ridges, one corresponding to each septum.   Upper calicular margin horizontal and entire.

Septa hexamerally arranged in five cycles. S_1 slightly larger than S_2, these slightly larger than S_3.   Septa of first three cycles not exsert and extending to columella.   Each of these septa bearing shallow, nondentate notch near its upper junction with theca.   Short thecal lip sometimes present, extending slightly above septal insertions. S_4 about half size of S_1, sometimes bending slightly toward S_3; S_5 rudimentary, extending about one fourth of way to base.   All septa with straight inner edges and covered by very low, rounded granules, usually arranged in rows parallel to septal edge.

Fossa shallow, containing prominent columella of variable structure. S_{1-3} often bearing one to three long paliform lobes, these sometimes round or flattened in cross section.   Four to thirty of these lobes may be present, forming columella. Sometimes columellar lamellae labyrinthine in arrangement (Plate 15, fig. 6) and sometimes greatly thickened (Plate 15, fig. 9).

Remarks.   Bullivant [1967] reported a deep slope cobble assemblage on the Pennell Bank, Ross Sea (461-583 m), characterized by G. antarctica, various echinoderms, sponges, bryozoans, and stylasterine coral.   Squires [1962b] had earlier reported G. antarctica, Flabellum antarcticum (actually F. flexuosum), and Caryophyllia

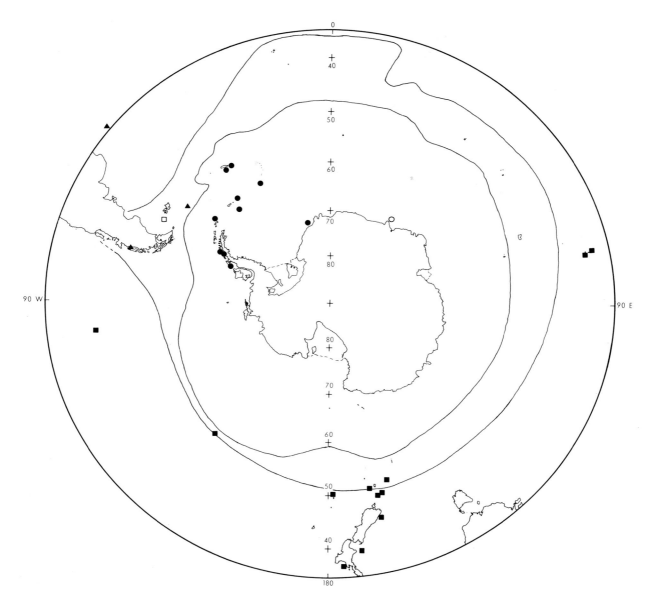

Map 12. Distribution of <u>Javania antarctica</u> (solid circles), Desmophyllum <u>pseudoseptatum</u> (open circle), <u>Javania cailleti</u> (solid triangles), Stenocyathus <u>vermiformis</u> (solid squares), and <u>Balanophyllia</u> sp. (open square).

<u>antarctica</u> from the same locality. These three species have also been collected together several times off Cape Hallett, Ross Sea, at 342-433 m (<u>Eltanin</u> stations 1870, 1995, 1996; <u>Atka</u> station 23; and <u>Burton Island</u> station 3), often attached to the stylasterine coral <u>Errina fissurata</u> (Gray). In the assemblage from off Cape Hallett, both <u>F. flexuosum</u> and <u>C. antarctica</u> are small and often scolecoid; the <u>C. antarctica</u> are often missing pali, resembling <u>Cyathoceras</u>.

One of the specimens from <u>Eltanin</u> station 2097 had a full gastrovascular cavity of disarticulated and partially digested, small, shrimplike crustaceans. Judging from the size of the antennule scales, two species are represented, the larger one probably measuring 20 mm (B. Kensley, personal communication, 1979).

<u>Discussion</u>. The specimens reported as <u>G. lilliei</u> by Gardiner [1929a] are typical small <u>G.</u>

<u>antarctica</u>. His redescription of <u>G. lilliei</u> [Gardiner, 1939] is based on a suite of three or four species, as he suggested may have been the case: <u>Flabellum flexuosum</u>, <u>Caryophyllia eltaninae</u>, <u>Flabellum gardineri</u>, and <u>Flabellum</u> sp. Gardiner's [1929a] <u>Gardineria</u> sp., from off New Zealand, resembles <u>Caryophyllia antarctica</u> in its septal ornamentation and columella but lacks the distinctive pali.

<u>G. antarctica</u> superficially resembles <u>G. capensis</u> (Gardiner, 1904) from off South Africa but can be distinguished by its extra cycle of septa, inequality in size of S_{1-3}, and lack of exsert septa.

<u>G. antarctica</u> is the only scleractinian fossil known from the Antarctic continent [Squires, 1962b, p. 15], reported from various localities around McMurdo Sound, Ross Sea, from the Pliocene-Lower Pleistocene to the Pleistocene-Subrecent. Speden

[1962] noted that the fossil specimens have a much thicker theca. One of the fossil specimens mentioned by Gardiner [1929a] is illustrated by David and Priestley [1914, pl. 88, figs. 4, 5].

Material. Eltanin sta. 1081 (17), USNM 47194; sta. 1082 (1), USNM 47183; sta. 1088 (2), USNM 47186; sta. 1089 (4), USNM 47206; sta. 1870 (125), USNM 47209; sta. 1871 (1), USNM 47181; sta. 1922 (2), USNM 47540; sta. 1924 (1), USNM 47200; sta. 1931 (4), USNM 47182; sta. 1933 (2), USNM 47203; sta. 1944 (3), USNM 47188; sta. 2021 (6), USNM 47189; sta. 2022 (1), USNM 47204; sta. 2045 (1), USNM 47195; sta. 2072 (2), USNM 47198; sta. 2075 (1), USNM 47191; sta. 2082 (4), USNM 47248; sta. 2097 (1), USNM 47202; sta. 2099 (1), USNM 47192; sta. 2117 (30), USNM 47205; sta. 2119 (38), USNM 47249; sta. 2120 (3), USNM 47180; sta. 2124 (3), USNM 47190; sta. 2125 (1), USNM 47185; sta. 5761 (4), USNM 47187; sta. 5762 (1), USNM 47247; sta. 5765 (9), USNM 47208. Hero sta. 691-20 (1), USNM 47196; sta. 731-1844 (1), USNM 47199. Edisto sta. 20 (1), USNM 53406. Glacier sta. 1 (1), USNM 47206. Staten Island sta. 21 (15), USNM 47193. EW sta. 9 (5), USNM 45663; sta. 32 (10), USNM 47197. NZOI sta. A-625 (4), USNM 47201. EAD sta. 2 (8), USNM 53399; sta. 3 (4), USNM 47184. Specimens (3) identified as Caryophyllia inskipi by Thomson and Rennet [1931], Australian Museum G 13537; specimens (2) identified as Gardineria antarctica from Discovery sta. 152 by Gardiner [1939], BM 1939.7.22.242-243. Syntypes of G. antarctica and G. lilliei.

Types. The seven syntypes of G. antarctica, collected at Terra Nova stations 314 and 349, are deposited at the British Museum. Type-locality: The six syntypes of G. lilliei are a mixed lot, represented by five specimens of G. antarctica and one basal fragment of a Flabellum, perhaps F. flexuosum. Therefore, one specimen (Plate 15, fig. 10) is designated as lectotype. All specimens are deposited at the British Museum (1929.10.22.9-14). Type-locality: 69°43'S, 163°24'E (off Oates Coast, Antarctica); 329-366 m.

Distribution. Circumpolar: off Antarctica; off South Shetland Islands; east of South Orkney Islands; off Scott Island (Map 11). Depth range: 87-728 m.

Family GUYNIIDAE Hickson, 1910
Genus Stenocyathus Pourtalès, 1871

Diagnosis. Solitary, ceratoid to cylindrical, free or attached. Wall epithecal; rows of thecal spots flank each S3. Pali, when present, opposite S2. Columella formed of one to four twisted, crispate ribbons. Type-species: Coenocyathus vermiformis Pourtalès, 1868, by monotypy.

34. Stenocyathus vermiformis (Pourtalès, 1868)
Plate 16, figs. 8-11

Coenocyathus vermiformis Pourtalès, 1868, pp. 133, 134.
Stenocyathus vermiformis; von Marenzeller, 1904a, pp. 298-300, pl. 18, fig. 16.--Zibrowius, 1974a, pp. 769, 770; 1980, pp. 163-165, pl. 84, figs. A-Q.--Cairns, 1979, pp. 168-170, pl. 32, figs. 8-10, pl. 33, figs. 1, 2.
Stenocyathus decamera Ralph and Squires, 1962, pp. 11, 12, pl. 4, figs. 2-6.--Squires and Keyes, 1967, p. 28, pl. 6, figs. 3-5.--Squires, 1969, p. 17, pl. 6, map 2.

Description. This species has been fully described and illustrated elsewhere [Cairns, 1979; Zibrowius, 1980]; only a brief description follows. Corallum cylindrical, elongate, vermiform, up to 50 mm long but rarely over 5 mm in CD. Free or attached; when attached, reinforced basally by layers of granular stereome. Epitheca thin, porcelaneous; usually marked by 24 longitudinal rows of white spots, 1 row corresponding to each interseptal space. Septa hexamerally arranged in three systems. S1 largest septa; S2 usually larger than S3 but may be of same size or smaller. Septa usually not exsert. Inner edges of septa sinuous. Thick pali before S2; columella formed of one to four twisted ribbons. Septa and pali bear large granules.

Discussion. Additional variations noted in the Subantarctic specimens include the following: (1) the S1 are sometimes exsert, (2) the S3 are sometimes equal to or larger than the S2, and (3) the columella may be composed of up to four elements.

The white spots on the theca are solid, smooth structures but probably represent areas of lesser calcification, as is evidenced by their earlier erosion to pores after death of the coral.

Material. Eltanin sta. 1284 (8), USNM 47450; sta. 1411 (1), USNM 47448; sta. 1691 (1), USNM 47454; sta. 1851 (1), USNM 47446. NZOI sta. A-740 (1), USNM 47449; sta. D-159 (9), USNM 47452; sta. D-160 (1), USNM 47451; sta. D-175 (1), USNM 47453. Specimens (9) identified as S. decamera by Squires and Keyes [1967] from NZOI sta. B-319, USNM 47447; specimens listed by Cairns [1979], USNM. Syntypes of C. vermiformis.

Types. The types of C. vermiformis are at the Museum of Comparative Zoology [see Cairns, 1979]. Type-locality: Florida Keys; 274-329 m. The holotype of S. decamera is deposited at the New Zealand Geological Survey, Wellington. Type-locality: off New Zealand; 110-220 m.

Distribution. Widely distributed in Atlantic Ocean: Mediterranean Sea; area bordered by Celtic Sea, Azores, and Madeira; western Atlantic from off Georgia, United States, to off Rio de Janeiro, Brazil; off Penedos de São Pedro e São Paulo (St. Peter and Paul Rocks). From more southern latitudes known from off Île Saint-Paul and Île Amsterdam, Indian Ocean; off New Zealand; Campbell Plateau; off Antipodes Islands; several seamounts in South Pacific (Map 12). Depth range: 80-1229 m.

Suborder DENDROPHYLLIINA Vaughan and Wells, 1943
Family DENDROPHYLLIIDAE Gray, 1847
Genus Balanophyllia Wood, 1844

Diagnosis. Solitary, turbinate to trochoid, fixed or free. Costae well developed. Synapticulotheca porous, especially near calicular edge. Septa follow Pourtalès plan. Pali present or absent. Columella spongy. Type-species: Balanophyllia calyculus Wood, 1844, by monotypy.

35. Balanophyllia malouinensis Squires, 1961
Plate 16, figs. 4-7; Plate 17, figs. 1-3
Plate 18, fig. 7

Balanophyllia cornu; Gardiner, 1939, pp. 335, 336.
Balanophyllia malouinensis Squires, 1961, pp. 15,

Plate 16. *Balanophyllia* and *Stenocyathus*

1-3.   *Balanophyllia* chnous Squires: holotype, BM 1929.10.22.25, Terra Nova sta.
       191, H = 30.4 mm, GCD = 12.1 mm.
4-7.   *Balanophyllia* malouinensis Squires: 4, specimen identified as B. cornu by
       Gardiner [1939], BM 1939.7.20.234, WS sta. 839, H = 34.4 mm; 5, 6, USNM
       45672, Eltanin sta. 558, GCD = 22.8 mm; 7, specimen from same lot, x1.8,
       coated with ammonium chloride to illustrate columella, stereotheca, and
       synapticulotheca.
8-11. *Stenocyathus* vermiformis (Pourtalès): 8, USNM 47447, NZOI sta. B-319, H =
       14.4 mm; 9, USNM 47454, Eltanin sta. 1691, H = 13.9 mm; 10, USNM 47466,
       Eltanin sta. 1851, CD = 4.0 mm; 11, USNM 47448, Eltanin sta. 1411, H = 7.4
       mm, illustrating exothecal deposits.

39, 40, 46, figs. 5, 24-26; 1969, pp. 17, 18, pl. 6, map 2.--Sorauf and Podoff, 1977, pp. 4-6, pl. 2, figs. 5, 6, pl. 3, fig. 6, pl. 4, figs. 2-5.--Cairns, 1979, p. 206.

Description. Corallum ceratoid to subcylindrical, straight to slightly curved, usually free when adult. Basal disc diameter about 4.2 mm, narrowing slightly to PD of about 3.5 mm, then gradually expanding into ceratoid corallum. Some coralla, however, subcylindrical and remaining firmly attached by strengthening base with layers of stereome, up to 12 mm in diameter. Young coralla usually weakly attached to epitheca of same species, F. curvatum, small gastropods, bivalves, or pebbles. Largest specimen examined 23.1 x 21.2 mm in CD and 57.5 mm tall. Synapticulotheca thick, very porous, and spinose. Costal spines usually randomly arranged, but in some specimens aligned longitudinally and separated by striae, resembling costae. Between 60 and 100% of synapticulotheca covered by thin, irregularly banded epitheca, often leaving only small ring of synapticulotheca visible at calicular edge. According to Sorauf and Podoff [1977], synapticulotheca gradually infilled by stereome, forming more solid 'stereotheca' (Plate 16, fig. 7).

Septa hexamerally arranged in five cycles but only largest specimens with complete fifth cycle. S_1 and S_2 equal in size and extending to columella. Remaining septa arranged in Pourtalès plan: S_4 smallest septa, S_5 adjacent to S_1 and S_2 larger than S_3 and extending to columella. Septa not exsert and with straight inner edges. Septal granulation variable, from very sparse (smooth septal faces) to crowded arrangement of tall, blunt granules.

Columella discrete, massive, elongate structure, resting in shallow, elliptical fossa; either spongy or composed of numerous twisted ribbons, swirled in clockwise direction. Columella may be granulated.

Discussion. Contrary to Squires's [1961] original description, based on six worn specimens, this species does not have exsert septa and does sometimes have costae. The presence or absence of costae is the basic difference between Balanophyllia and Thecopsammia, the latter lacking costae. The variable nature of this character in B. malouinensis implies that Thecopsammia may be a junior synonym of Balanophyllia. Contrary to Wells [1956] and Squires [1961], Thecopsammia socialis (type-species of Thecopsammia) has a distinct Pourtalès plan in the adult stage [see Cairns, 1979].

Material. Eltanin sta. 339 (255), USNM 47179; sta. 340 (15), USNM 45671; sta. 346 (1), USNM 47154; sta. 369 (6), USNM 47149; sta. 558 (221), USNM 45672; sta. 740 (103), USNM 47153; sta. 970 (21), USNM 47146; sta. 977 (205), USNM 47148; sta. 1521 (6), USNM 47147; sta. 1536 (12), USNM 47178; sta. 1596 (1), USNM 47155. Hero sta. 715-895 (7), USNM 47177. Vema sta. 14-12 (5), AMNH; sta. 14-18 (1), AMNH; sta. 15-PD3 (9), USNM 47150; sta. 15-PD4 (78), USNM 47151; sta. 15-PD9 (6), AMNH; sta. 15-PD10 (9), USNM 53408, and (24), AMNH; sta. 17-59 (6), USNM 53409, and (11), AMNH. Following WH records (H. Zibrowius, personal communication, 1979): sta. 324/66 (2), sta. 325/66 (1), sta. 330/66 (1), sta. 336/66 (3), sta. 357/66 (2), sta. 359/66 (2), sta. 360/66 (1), sta. 269/71 (4), sta.

270/71 (1) (all WH specimens desposited at ZIZM). Specimens (5) identified as B. cornu by Gardiner [1939], BM 1939.7.20.227-228, 234. Holotype.

Types. The holotype is deposited at the American Museum of Natural History (3368). Type-locality: 52°32'S, 61°15'W (south of East Falkland island); 358 m.

Distribution. Off Tierra del Fuego; off Falkland Islands; Scotia Ridge between Burdwood Bank and South Georgia (Map 13). Squires's [1969] record from off Gough Island unsubstantiated. Depth range: 75-1137 m.

36. Balanophyllia sp.
Plate 17, figs. 4-8

Dendrophyllia oahensis; Gardiner, 1939, pp. 334, 335.--Squires, 1961, p. 21.

Description. In the following, William Scoresby station 244 specimens are described. Ceratoid to cylindrical coralla, attached to dead coralla of same species in pseudocolonial arrangement. One 'corallite' 34.4 mm long and cylindrical (CD = 7.7 x 5.7 mm); others ceratoid, shorter, with larger calices (GCD up to 14.0 mm). Thin epitheca covers most of porous synapticulotheca; synapticulotheca longitudinally striate. Septa hexamerally arranged in four complete cycles with no S_5. S_1 and S_2 equal in size and extending to columella. S_3 half as large and enclosed by pairs of larger S_4, these sometimes fusing near inner edge of S_3 and extending to columella. Septa not exsert, slightly porous, especially toward theca, and bearing large, pointed granules. Columella large and spongy.

Discussion. These specimens were collected in the geographic and depth range of B. malouinensis but are distinguished by their cylindrical shape, tendency toward pseudocoloniality, and lack of S_5, even at a relatively large calicular diameter. They are, however, similar to small colonies of the northern Atlantic Dendrophyllia cornucopia Pourtalès, 1871. D. cornucopia sometimes has extratentacular budding similar to that of the pseudocolonial Balanophyllia sp., and the septal arrangement is identical when small corallites of D. cornucopia are considered.

The holotype of D. oahensis Vaughan, 1907, known only from off Hawaii, differs from these specimens by having sinuous septal edges and a truly colonial growth form.

Material. WS sta. 244 (3 pseudocolonies), BM, and (1 pseudocolony), MCZ 3571.

Distribution. Known only from 52°00'S, 62°40'W (east of East Falkland island) (Map 12). Depth range: 247-253 m.

37. Balanophyllia chnous Squires, 1962
Plate 16, figs. 1-3

Thecopsammia sp. Gardiner, 1929a, pp. 126, 127.
Balanophyllia sp. Gardiner, 1929a, pp. 126, 127.
Balanophyllia chnous Squires, 1962b, pp. 13, 21, 22, pl. 1, fig. 17, pl. 2, figs. 1-3; 1969, pp. 17, 18, pl. 6, map 2.

Description. In the following the holotype is redescribed. Corallum ceratoid, becoming cylindrical toward calice, firmly attached to substrate by pedicel 6.2 mm in diameter. Calice elliptical,

Plate 17. Balanophyllia

1-3. Balanophyllia malouinensis Squires: 1, USNM 47179, Eltanin sta. 339, x2.4; 2, USNM 47153, Eltanin sta. 740, H = 19.1 mm, coated with ammonium chloride; 3, USNM 47179, Eltanin sta. 339, x1.9, 3 young, attached specimens.

4-8. Balanophyllia sp. (specimens identified as B. oahensis by Gardiner [1939], BM (no number), WS sta. 244): 4, GCD = 11.1 mm; 5, GCD = 13.2 mm; 6, GCD = 7.7 mm; 7, x2.0; 8, x2.2

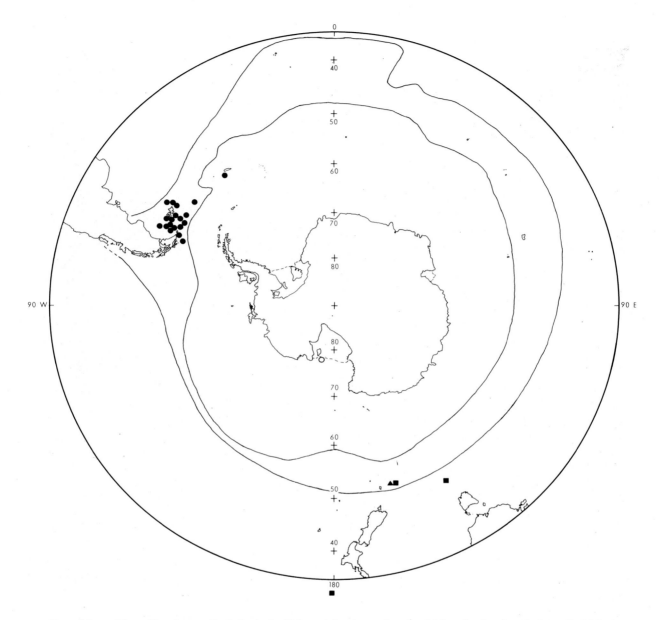

Map 13.   Distribution of <u>Balanophyllia</u> <u>malouinensis</u> (solid circles), <u>Balanophyllia</u> <u>chnous</u> (open circle), <u>Enallopsammia</u> <u>rostrata</u> (solid squares), and <u>Enallopsammia</u> <u>maren-</u><u>zelleri</u> (solid triangle).

12.1 x 9.3 mm in diameter; corallum 30.4 mm tall. Epitheca covering porous synapticulotheca to within 3 mm of calice. Epitheca heavily encrusted with Bryozoa, Foraminifera, serpulids, and sponges. Costae not apparent but may have been present before epitheca formed. Sixty-six septa hexamerally arranged in five incomplete cycles. S_1 largest septa, slightly exsert, and extending to columella. S_2 slightly smaller, not reaching columella. Remaining septa arranged in Pourtalès plan, those adjacent to S_1 and S_2 large and usually extending to columella. Development of S_5 irregular within systems: the two lateral systems without S_5, but the four systems adjacent to lateral edges with variable number of S_5. Three half systems with all four S_5. Inner edges of $S_{1,2}$ straight and entire; those of S_{3-5} irregular to laciniate. Septal granules prominent, as large as septal thickness in height. Short,

low carinae present on inner edges of larger septa. Columella elongate and spongy.

Discussion. <u>B</u>. <u>chnous</u> is the only dendrophylliid known from off continental Antarctica. The only specimens known were reported by Gardiner [1929a] as <u>Thecopsammia</u> sp. and <u>Balanophyllia</u> sp. He stated that the specimens were collected at <u>Terra</u> <u>Nova</u> station 91 (off New Zealand) at the beginning of his species account and at <u>Terra</u> <u>Nova</u> station 191 (Bay of Whales, Ross Sea) at the end of the species account. The original label reads 191. However, Gardiner also validly reported specimens of other species from <u>Terra</u> <u>Nova</u> station 91 in the same paper. If there was a labeling error, it might explain the subsequent lack of records of this species in the Ross Sea, even though several <u>Eltanin</u> stations have been made in the proximity of the type-locality.

<u>B</u>. <u>chnous</u> differs from <u>B</u>. <u>malouinensis</u> by its

exsert septa; smaller size at maturity; and longer, more slender corallum.

Material. Holotype.

Types. The holotype (number 1929.10.22.25) and five paratypes (numbers 1929.10.22.22-24, 26-27) are deposited at the British Museum. They were presumably collected at Terra Nova station 191. Type-locality: Bay of Whales, Ross Sea; 355-457 m.

Distribution. Known only from type-locality (Map 13).

Genus Enallopsammia Michelotti, 1871

Diagnosis. Dendroid (often uniplanar) colonies formed by extratentacular budding. Coenosteum compact, synapticulothecate, porous only near calices. Septa arranged normally. Columella small. Type-species: Coenopsammia scillae Seguenza, 1864, by monotypy.

38. Enallopsammia rostrata (Pourtalès, 1878)
Plate 18, figs. 1-4

Amphihelia rostrata Pourtalès, 1878, p. 204, pl. 1, figs. 4, 5.
Stereopsammia rostrata; Pourtalès, 1880, pp. 97, 110, 111.
Dendrophyllia (Coenopsammia) amphelioides Alcock, 1902, pp. 42, 43, pl. 5, figs. 37, 37a.--Not D. (C.) amphelioides; Gardiner and Waugh, 1939, p. 238.
Amphihelia adminicularis Rehberg, 1892, p. 10, pl. 4, fig. 1.
Anisopsammia rostrata; von Marenzeller, 1904a, pp. 314, 315, pl. 18, fig. 23.--Gravier, 1920, pp. 102-104, pl. 12, figs. 181-185.
Anisopsammia amphelioides; Vaughan, 1907, pp. 156, 157, pl. 47, figs. 1, 2.
Dendrophyllia amphelioides var. cucullata Vaughan, 1907, p. 157, pl. 47, fig. 3, pl. 48, figs. 1-4.
Madrepora ramea; Gardiner and Waugh, 1939, pp. 226, 227.
Dendrophyllia minuscula; Gardiner and Waugh, 1939, p. 237 (part).
Enallopsammia rostrata; Squires, 1959, p. 40.--Laborel, 1970, p. 156.--Zibrowius, 1973, pp. 44, 45, pl. 2, figs. 14, 15; 1980, pp. 201-203, pl. 105, figs. A-K, pl. 106, figs. A-C.--Cairns, 1979, pp. 186-188, pl. 37, figs. 2, 3, 6.
Enallopsammia amphelioides; Zibrowius, 1973, pp. 45-48, pl. 3, figs. 16-20; 1980, pp. 203, 204, pl. 106, figs. D-I.

Description. Colony flabellate, dendroid; extratentacular branching occurring at every or every second or third corallite. Base of colony massive, up to 3 cm in diameter. Calices 3-5 mm in diameter, round to teardrop shaped, occurring on only one side of colony. Calices projecting upward costoseptal rostrum, formed by enlargement of one CS_1 and sometimes adjacent septa. Rostrum variable in development and sometimes absent. Both sides of branches costate, coenosteum solid. Septa hexamerally arranged in three cycles, rarely with additional S_4. S_1 largest septa, septa of other two cycles progressively smaller; S_3 sometimes fused to S_2 halfway to columella. Septa not exsert except for S_1 forming rostrum. Inner edges of septa variable, from smooth to laciniate, sometimes bearing wide paliform lobes. All septa narrow, thick near calice and thinner toward

columella, and bearing spiny granules. Columella rudimentary, spongy.

Discussion. More complete descriptions, additional illustrations, and an explanation of the synonymy can be found in works by Zibrowius [1973, 1980] and Cairns [1979].

According to Zibrowius [1973] the main difference between E. rostrata and E. amphelioides is the degree of development of the exsert S_1, or rostrum: the former has a prominent rostrum, whereas the latter has a reduced rostrum or none at all. After reexamination of Atlantic specimens previously reported [Cairns, 1979] and additional specimens from the eastern Atlantic, off Hawaii, and off New Zealand, I am led to agree with Vaughan [1907] that there is a continuous intergradation between the extreme rostrate and nonrostrate forms, not only off Hawaii but also in the western Atlantic and off New Zealand. In fact, a single branch from an Albatross station 3827 specimen bears extremely rostrate calices, nonrostrate calices, and some that are intermediate in development with only a slightly developed S_1. The development of the rostrum seems to be influenced by microenvironmental changes and may be a reaction to a poor feeding area. The most extreme rostra are often those of calices which are adjacent to other corals or in close proximity to other corallites of the same colony. I agree with Vaughan [1907] that these differences should be designated as varietal or formae, not as separate species. It may be useful to refer to all specimens with a rostrum, or enlarged S_1, as the typical form and to those without enlarged S_1 as forma amphelioides, because some colonies have calices that uniformly lack rostra, whereas calices of other colonies all have rostra.

E. rostra is differentiated from other Enallopsammia by its unifacial calices.

Material. Eltanin sta. 1411, USNM 47531; sta. 1981, USNM 47532; sta. 1983, USNM 47533. NZOI sta. C-527, USNM 47534. Specimens listed by Cairns [1979], USNM. Syntypes of A. rostrata; syntypes of Dendrophyllia amphelioides var. cucullata.

Types. The syntypes of A. rostrata are deposited at the Museum of Comparative Zoology. Type-locality: 23°14'N, 82°25'W (Straits of Florida); 1472 m. The syntypes of D. (C.) amphelioides are deposited at the Zoölogische Museum, Amsterdam. Type-locality: off Pulau Waigeo and Pulau Misool, Indonesia; 469-1633 m. The syntypes of D. amphelioides var. cucullata are deposited at the United States National Museum. Type-locality: off Hawaii; 426-679 m.

Distribution. Widely distributed except for eastern Pacific: western Atlantic (San Pablo Seamount to Río de Janeiro, Brazil), eastern Atlantic (area bounded by Celtic Sea, Azores, and Gulf of Guinea), Indian Ocean (off Maldive, off Nicobar Islands), western Pacific (off Japan, Indonesia), central Pacific (off Hawaii, Îles Tuamotu), South Tasmanian Rise, Macquarie Ridge, Kermadec Ridge (Map 13). Depth range: 229-2165 m.

39. Enallopsammia sp. cf. marenzelleri
Zibrowius, 1973
Plate 18, figs. 5, 6

Enallopsammia marenzelleri Zibrowius, 1973, pp. 49-51, pl. 1, figs. 1-7, pl. 2, figs. 8-11; 1980, pp. 204, 205, pl. 106, figs. J-M.

Plate 18. Enallopsammia, Balanophyllia, and Caryophyllia

1-4.   Enallopsammia rostrata (Pourtalès): 1, 2, USNM 47532, Eltanin sta. 1981,
       x0.85, dyed red; 3, 4, calices from other branches in same lot, both x5.1,
       coated with ammonium chloride.

5, 6. Enallopsammia marenzelleri Zibrowius: 5, USNM 47535, Eltanin sta. 1411,
       x5.5, dyed red; 6, same specimen, x1.8.

7.   Balanophyllia malouinensis Squires: USNM 47179, Eltanin sta. 339, H = 57.7
       mm.

8, 9. Specimen identified as Caryophyllia clavus var. smithi by Moseley [1881]:
       BM 1880.11.25.27, Challenger sta. 308, GCD = 8.7 mm.

Discussion. Thirteen branch fragments, the largest bearing only 13 corallites, are tentatively assigned to this species. Branching bushy, calices on alternate sides of branches. Distinct costae present. Calices round, about 4–5 mm in diameter, without enlarged S_1, or rostra. Otherwise, aspects of septa and columella similar to those of previously described species.

E. marenzelleri is a poorly known species, originally described from only five colonies, two of which are equal to or smaller than these fragments. On the basis of the original description these fragments seem to belong to E. marenzelleri, but not enough material is present to be conclusive. This species is distinguished from the other two Recent species with alternately placed calices (E. pusilla, E. profunda) by its possession of costae.

Material. Eltanin sta. 1411 (13 fragments), USNM 47535.

Types. The holotype and paratype of E. marenzelleri are deposited at the Zoölogische Museum, Amsterdam (Coel. 6902, 588). Type-locality: 5°56.5'S, 132°47.7'E (off Kepulauan Kai (Kei Islands)); 595 m.

Distribution. Meteor Seamount; off Azores; off Nicobar Islands; off Kepulauan Kai (Kei Islands); Macquarie Ridge (Map 13). Depth range: 371–815 m.

Uncertain Records

Caryophyllia clavus var. smithi Broderip, 1828
Plate 18, figs. 8, 9

Caryophyllia clavus var. smithi; Moseley, 1881, p. 134 (part: Challenger sta. 308).

Moseley reported this species from off southern Chile (Map 3) on the basis of three specimens, two of which were damaged (BM 1880.11.25.27). Complete specimen firmly attached to gorgonian stem and 8.7 x 7.6 mm in CD and 18.7 mm tall. Theca porcelaneous, marked by thin, vertical striae, which delimit costae. Costal granules low and rounded. Septa hexamerally arranged in four cycles but missing two S_4 for 46 septa. S_1 and S_2 equal in size and moderately exsert; S_3 and S_4 progressively smaller and less exsert. Septal edges straight except for those of S_3. Crown of eleven P_3 enclosing elongate columella composed of eight narrow, twisted ribbons.

These species are clearly not C. clavus var. smithi sensu Duncan, 1876 (actually C. smithi Broderip, 1828), thus far known only from the eastern Atlantic from Norway to the Congo [Zibrowius, 1980]. They also do not belong to any of the Caryophyllia discussed in this text. They might represent an undescribed species, but more specimens are needed for a complete description.

Flabellum transversale conicum
Yabe and Eguchi, 1942

Flabellum transversale conicum; Eguchi, 1965, p. 11, pl. 1, figs. 3a, 3b.

Eguchi reported one specimen from off Riiser-Larsen Peninsula (Cape Cook), Antarctica, at 920 m. Unfortunately, the specimen is not available for study. Judging from the illustration provided by Eguchi, the specimen could be a basal fragment of F. flexuosum or F. impensum, both of which are known from this region.

Flabellum ongulense Eguchi, 1965

Frabellum lurvatum; Niino, 1958, pl. 2, fig. 2 (misspellings).
Flabellum ongulense Eguchi, 1965, pp. 11, 12, pl. 2, figs. 2a–2d.

Eguchi described this species on the basis of one specimen from off Riiser-Larsen Peninsula (Cape Cook), Antarctica, at 750 m (Map 10). Because the holotype is not available for study, it is not possible to add anything to his original description. It is very similar to F. flexuosum but difficult to verify from Eguchi's account.

Desmophyllum pseudoseptatum Eguchi, 1965

Desmophyllum delicatum; Niino, 1958, pl. 2, fig. 3.
Desmophyllum pseudoseptata [sic] Eguchi, 1965, p. 9, pl. 2, figs. 3a–3c.

Eguchi described this species on the basis of one specimen from off Riiser-Larsen Peninsula (Cape Cook), Antarctica, at 630–680 m (Map 12). The holotype is not available for study, but Eguchi's illustrations and description very much resemble those of Javania antarctica (Gravier, 1914). Eguchi apparently intended to name this species D. pseudocostatum.

Zoogeographic Analysis

The Antarctic and Subantarctic regions have not been uniformly and thoroughly sampled, especially not the region from the Antarctic Peninsula to the Ross Sea and from the Ross Sea to the Weddell Sea. Furthermore, several scleractinian species from this area are known from only one or two records, some are doubtfully assigned, and records of several unique specimens have not been discussed in this paper pending the collection of additional specimens. Nonetheless, it is possible to make some generalizations about patterns of distribution and regional affinities on the basis of the USARP specimens and a reevaluation of previously reported specimens.

For conformity, the zoogeographic divisions and terminology of Hedgpeth [1969] will be used in this paper (Map 14). The Antarctic region is the area inside the Antarctic convergence, including South Georgia, Bouvetøya, and Heard Island. The Subantarctic boundary follows the subtropical convergence only partially and includes Magellanic South America, Falkland Islands, Tristan da Cunha Group, Gough Island, Prince Edward Islands, Îles Crozet, Îles Kerguelen, Macquarie Ridge, Campbell and Auckland islands, and several unnamed seamounts in the South Pacific. Areas directly to the north of the Subantarctic are referred to as cold temperate.

Patterns of Distribution

Among the 37 species of Scleractinia known from the Antarctic-Subantarctic region, certain patterns of distribution occur (Table 2, last column), most

Map 14. Locator map indicating boundaries of Antarctic and Subantarctic regions [after Hedgpeth, 1969].

of which are shared with one other group or more of benthic invertebrates. The patterns are as follows:

I. Antarctic region species
    A. Circumpolar or widespread in region, usually including Scotia Ridge but not crossing Antarctic convergence.
    B. Endemic to single localities.
II. Subantarctic region species
    A. Primarily Magellanic.
    B. Campbell Plateau, Macquarie Ridge, and usually New Zealand and/or Australia.
    C. Circum-Subantarctic.
    D. Endemic to a seamount.
III. Cosmopolitan or widespread species (with southern limit as given below)
    A. Subantarctic, but only marginally.
    B. Antarctic region, insular or seamounts.
    C. Antarctic region, continental.

Eight species are endemic to the Antarctic region, a pattern found in most other benthic invertebrate groups. Two of these, _Caryophyllia eltaninae_ and _Flabellum gardineri_, are only known from the South Georgia-Shag Rocks area, and _Balanophyllia chnous_ is known from only one record in the Ross Sea (pattern IB). Of the remaining five (pattern IA), four are circumpolar, including _Caryophyllia antarctica_, representing the ideal Antarctic distribution, including Bouvetøya. The fifth species, _Javania antarctica_, is found primarily off western Antarctica and the Scotia Ridge.

One of the circumpolar species, _Flabellum impensum_, was collected once north of the Antarctic convergence off the Antipodes Islands (_Eltanin_ station 2143) at its second greatest recorded depth. This record is not so unexpected if one considers that the distribution of benthic shelf and slope animals does not necessarily follow the zoogeographic boundaries imposed by surface currents [Squires, 1946c, p. 454; Kussakin as cited by Dell, 1972, p. 11]. Instead, the characteristics of the water mass must be considered, which could be Antarctic bottom water or Antarctic intermediate water off the Antipodes Islands, depending on depth [Menzies et al., 1973, p. 201]. The temperatures for the Antarctic records of F. impensum are probably at or below 0°C, fully within the Antarctic bottom water mass. Ridgway [1968] indicated a bottom temperature of 3°-4°C for the Antipodes Islands locality, but it is an area of highly compressed benthic isotherms over short distances. Upwelling may have brought colder Antarctic bottom water closer to the surface here, or the species may have adapted to slightly warmer water at this latitude. Dawson [1970] has reported similar distributions for primarily Antarctic Asteroidea and Ophiuroidea, hypothesizing a possible migration route from the Ross Sea to the Balleny Islands, then along the ridge system from the Balleny Islands to Macquarie Island, and then to the Campbell Plateau and New Zealand. Dawson concluded that this migration route is probably not a common one but that it may explain some of the echinoderm distributions; the Antipodes Islands record of F. impensum may result from a similar migration.

Nine species are characteristic of the Magellanic subregion (pattern IIA), two of which are known from only single records (Phyllangia fuegoensis and Balanophyllia sp.). Of the remainder, five are restricted to the eastern coast of South America. Of these, two cross over the Antarctic convergence to South Georgia, and three extend northward as far as Río de la Plata. Sphenotrochus gardineri occurs on both sides of South America. The ninth species, Flabellum truncum, has been collected in the Magellanic subregion but also extends up the western coast as far as Peru. Two other species should be mentioned here: Bathelia candida is more characteristic of the cold temperate region to the north but does extend into the Magellanic subregion, off both the east and the west coast of South America. A. rathbuni is known from the warm temperate and subtropical coasts of eastern South America and is known from the Subantarctic only as a subfossil at Tierra del Fuego.

Three species (pattern IIB) are known from the Subantarctic islands of the Campbell Plateau and/or off Macquarie Island as southernmost records of species normally found in more northerly, cold temperate regions, such as off New Zealand, the Chatham Rise, and off southern Australia. One species, Flabellum apertum (pattern IIC), is essentially circum-Subantarctic, and Cyathoceras irregularis is endemic to a Subantarctic seamount (pattern IID).

Another large component, including 11 species, does not fit any pattern but represents southernmost extensions of cosmopolitan or widespread species. Six of these species penetrate the Subantarctic but do not cross the Antarctic convergence (pattern IIIA). Of these, 5 occur off the Subantarctic islands south of New Zealand, and 1 (J. cailleti) is found in the western Magellanic subregion. Five cosmopolitan species cross the Antarctic convergence, 3 of them extending as far south as the Antarctic islands and seamounts (pattern IIIB) and 2 reaching the coast of Antarctica (pattern IIIC). One of the last group

(Leptopenus discus), however, might be considered an abyssal instead of a continental Antarctic species, its shallowest record being 2000 m.

The remaining two species have unusual and disjunct distributions. C. profunda is known from the southern cold temperate regions; from Subantarctic Tristan and Gough islands; and from one record from off Hugo Island, Palmer Archipelago. C. mabahithi is known from the northern Indian Ocean and, like C. profunda, from one record off the Palmer Archipelago.

A peculiarity of the Antarctic scleractinian distributions is the absence of the pattern common to most benthic invertebrate groups; that is, that species are distributed throughout, but endemic to, the Antarctic-Subantarctic region [see Dell, 1972, fig. 3]. The Antarctic convergence, despite the fact that intermediate and bottom waters cross freely and are little changed below it, does seem to form a boundary for many scleractinian distributions. The only exceptions are (1) C. profunda, (2) C. mabahithi, (3) four cosmopolitan species, (4) one Antarctic species (F. impensum) which crosses to the north, and (5) two Magellanic species which cross over to South Georgia. Nonetheless, the typical Antarctic-Subantarctic benthic invertebrate pattern is not achieved by any scleractinian.

Another characteristic of the Antarctic Scleractinia is their unusually high percentage of cosmopolitan species. Even if the abyssal Leptopenus discus is not included, 3 cosmopolitan species (8% of the Antarctic-Subantarctic species) occur in the Antarctic region, and 11 (31%) occur as far as the Subantarctic region. This is considerably more than is the case for other benthic invertebrate groups.

The group of Antarctic benthic invertebrates that most closely resembles the Scleractinia in distributional patterns is the Echinoidea [see Pawson, 1969]. This class has most of the distributional patterns described for Scleractinia and is one of the few groups with a limited number of species that cross the Antarctic convergence. The position of South Georgia as a transitional area is similar for both groups.

Geographic Affinities

Antarctic. Seventeen species of Scleractinia are known from the Antarctic region: eight endemic, five cosmopolitan, and two primarily Magellanic ones (which cross over to South Georgia) and two species with anomalous distributions (Caryophyllia profunda and C. mabahithi). The percent of endemic species is therefore 47%, close to Squires's [1969, p. 17] value of 50%, despite numerous additions and synonymies. (For this comparison, Squires's Caryophyllia A is included as a nonendemic Antarctic species.) If the percent of endemic species is calculated in accordance with Cailleux's [1961] method, which would exclude the two Magellanic crossovers, this figure would increase to 53%.

The scleractinian fauna of South Georgia and Shag Rocks (eight species) is a faunal mixture of Antarctic, Magellanic, endemic, and cosmopolitan species, typical of most other groups of benthic invertebrates [Dell, 1972, p. 7] and indicative of the mixed hydrologic conditions at these islands.

TABLE 2. Geographic Distribution of Antarctic and Subantarctic Scleractinia

Species Discussed in This Study	Western South America (35°–45°S)	Eastern South America (35°–45°S)	Western South America (45°–55°S)[a]	Eastern South America (45°–55°S)[a]	Falkland Islands, Burdwood Bank[a]	Tristan da Cunha Group and Gough Island	Prince Edward Islands	Macquarie Ridge[b]	Subantarctic Seamounts[c]	Campbell Plateau[d]	Bounty Platform[e]	New Zealand	Chatham Island and Chatham Rise	South Georgia, Shag Rocks	South Sandwich Islands	South Orkney Islands	South Shetland Islands	Western Antarctic Peninsula to Ross Sea[f]	Ross Sea[g]	Ross Sea to Weddell Sea[h]	Weddell Sea and Eastern Antarctic Peninsula[i]	Bouvetøya	South Africa	Île Saint-Paul and Île Amsterdam	South Australia, Tasmania	Widespread	Depth (Worldwide)	Distribution Pattern[j]
1, F. marenzelleri	x				x													x		x	x				x	x	300–5870	IIIC
2, F. fragilis								x				x			x	x			x							x	285–2200	IIIA
3, L. discus															x	x										x	2000–3566	IIIC
4, A. rathbuni		x	x	x																							?–90	k
5, P. fuegoensis			x	x																							?	IIA
6, B. candida			x	x																				x		x	500–1250	k
7, M. oculata									x			x											x	x		x	80–1500	IIIB
8, C. antarctica														x	x	x	x	x	x	x	x	x					87–1435	IA
9, C. squiresi				x				x				x	x														406–659	IIA
10, C. profunda														x									x	x			35–1116	k
11, C. eltaninae														x													101–778	IB
12, C. mabahithi																								x	x		278–1022	k
13, C. irregularis									x									x									549	IID
14, Cyathoceras A													x														2305–2329	1
15, S. platypus								x																	x		622–913	1
16, A. recidivus											x	x															128–732	IIB
17, S. gardineri	x	x												x									x				9–403	IIA
18, D. cristagalli					x	x		x	x	x	x	x	x	x									x	x	x	x	35–2460	IIIB
19, L. prolifera	x					x		x	x	x	x	x	x										x	x	x	x	60–2170	IIIA

	Depth (m)	Zone
20, S. variabilis	220–2165	IIIB
21, G. dumosa	100–638	IIB
22, F. thouarsii	71–305	IIA
23, F. areum	647–2229	IIA
24, F. curvatum	115–1137	IIA
25, F. impensum	46–2260	IA[m]
26, F. flexuosum	101–659	IA
27, F. gardineri	177	IB
28, F. knoxi	201–914	IIB
29, F. apertum	220–1500	IIC
30, F. truncum	595–1896	IIA
31, J. cailleti	86–2165	IIIA
32, J. antarctica	53–1280	IA
33, G. antarctica	87–728	IA
34, S. vermiformis	80–1229	IIIA
35, B. malouinensis	75–1137	IIA
36, Balanophyllia sp.	247–253	IIA
37, B. chnous	355–457	IB
38, E. rostrata	229–2165	IIIA
39, E. marenzelleri	371–815	IIIA

Total number of occurrences: 3 7 5 10 11 3 2 11 5 4 9 4 8 5 8 6 6 1 4 6 7 11

Underlined numerals indicate that more species are known from these areas but that they are not represented in the Antarctic-Subantarctic regions.

a Magellanic subregion.
b Hjort Seamount, Macquarie Island, and grounds between Macquarie Island and South Island, New Zealand [see Brodie and Dawson, 1965]. Eltanin stations 1411, 1412, 1414, 1422; NZOI stations D-6, D-159, D-166, C-734.
c Eltanin stations 1344–1346, 1691.
d Including Auckland Island and Campbell Island.
e Including Antipodes Islands and Bounty Islands (not Subantarctic). Eltanin stations 1851, 2143; NZOI station A-706.
f 55°W to 140°W, including offshore islands.
g 140°W to 170°E.
h 170°E to 10°W.
i 10°W to 55°W.
j See text (p. 60) for key.
k Distribution pattern not easily categorized. See text for details.
l Not found in Antarctic-Subantarctic regions.
m Including one record from off the Antipodes Islands.

South Georgia is fully within the Antarctic region and therefore 46°–50°C colder than Magellanic South America but several degrees warmer than the other islands on the Scotia Ridge [Briggs, 1974]. It is surrounded by vigorous upwelling and is fully within the West Wind drift, east of the Falkland Islands. Of the eight species that are known from off South Georgia, three are Antarctic in affinity, two are endemic, one is cosmopolitan, and two are Magellanic species that have crossed over the Antarctic convergence.

Only one scleractinian is known from off Bouvetøya, Caryophyllia antarctica. None are known from off Heard Island.

Subantarctic. The characteristic species of the Subantarctic Magellanic subregion have been discussed previously; however, when the cosmopolitan species are added and the area is subdivided, the following totals result: Magellanic western South America, 5 species; Magellanic eastern South America, 10 species (including two subfossils); and Magellanic region-Falkland Islands-Burdwood Bank, 11 species. Even when the subfossils are excluded, the eastern coast of South America has more known species than the western coast. The rich fauna of the Falkland Islands-Burdwood Bank area consists of 6 species characteristic of the Magellanic subregion, 4 cosmopolitan species, and 1 endemic species (Balanophyllia sp.).

Only three species have been reported from the Subantarctic district of the Tristan da Cunha Group: two cosmopolitan species and Caryophyllia profunda, which is basically a cold temperate species.

Only two species of Scleractinia are known from off the Prince Edward Islands: Solenosmilia variabilis (cosmopolitan) and Flabellum apertum (circum-Subantarctic). No species are known from the shelf or slopes off Îles Crozet, but two cosmopolitan species, Leptopenus discus and Fungiacyathus marenzelleri, were collected west of the islands at 2926 m. No Scleractinia have been reported from off Îles Kerguelen.

Eleven species are known from off Macquarie Island, Hjort Seamount, and the Subantarctic Ridge north of Macquarie Island. Eight of these species are cosmopolitan, one is circum-Subantarctic, and two are primarily characteristic of New Zealand and southern Australia with southern ranges in the Subantarctic. Although the Simpson Index (number of species common to both areas times 100 per number of species in smaller fauna) for this region and the Prince Edward Islands is 100 and for this region and the Subantarctic seamounts 80, these high affinities reflect an influence of cosmopolitan species, not a distinctive fauna that might be characterized as 'Kerguelen' [sensu Hedgpeth, 1969, fig. 10]. Only four species are known from the closely adjacent Campbell Plateau, all of which are also found off New Zealand. Two are cosmopolitan in distribution and two are characteristic of the New Zealand area.

Five species are known from the Subantarctic seamount or ridge on the Heezen fracture zone of the Eltanin fracture zone system (Eltanin stations 1344-1346): three are cosmopolitan, one is endemic, and one is unidentified (Caryophyllia sp.). A sixth species, the cosmopolitan Stenocyathus vermiformis, was also collected from a different Subantarctic Pacific seamount (Eltanin station 1691). According to Knox [1970], very little is known about the fauna of the seamounts and guyots of the South Pacific. Scleractinia are herein reported from six seamounts or high areas, including the two Subantarctic ones previously mentioned, one in the Antarctic region (Eltanin stations 254-255), and three north of the Subantarctic region (Eltanin stations 1718, 1284, and 326). The fauna at Eltanin stations 1344-1346 is extremely abundant and diverse, as was previously discussed (see account of Solenosmilia variabilis). There is a fairly characteristic fauna of four or five cosmopolitan species inhabiting the seamounts and islands of the Subantarctic region, including the Hjort Seamount, Macquarie Island, and off Prince Edward Islands. These species are D. cristagalli, M. oculata, S. variabilis, S. vermiformis, and F. apertum.

Appendix
TABLE A1. Station List

Station	Latitude	Longitude	Depth, m	Date
		USNS Eltanin		
13	56°20'N	51°58'W	3488-3492	March 29, 1962
18	58°15'N	48°36'W	3404-3422	April 3, 1962
138	62°00'S	61°09'W	1437	Aug. 8, 1962
214	42°07'S	74°32'W	145	Sept. 13, 1962
216	52°53'S	75°36'W	1190-1263	Sept. 16, 1962
217	54°22'S	64°42'W	106-110	Sept. 23, 1962
254	59°49'S	68°52'W	512-622	Oct. 10, 1962
255	59°44'S	68°51'W	1043-1208	Oct. 10, 1962
272	64°54'S	68°21'W	412	Oct. 21, 1962
338	53°09'S	59°37'W	587-595	Dec. 3, 1962
339	53°05'S	59°31'W	512-586	Dec. 3, 1962
340	53°08'S	53°23'W	567-578	Dec. 3, 1962
346	54°02'S	58°42'W	101-119	Dec. 4, 1962
353	55°15'S	58°55'W	3514-3642	Dec. 5, 1962
369	54°04'S	63°35'W	247-293	Dec. 12, 1962
370	53°54'S	64°36'W	104-115	Dec. 12, 1962
410	61°18'S	56°09'W	220-240	Dec. 31, 1962

TABLE A1. (continued)

Station	Latitude	Longitude	Depth, m	Date
		USNS Eltanin (continued)		
416	62°40'S	56°13'W	494-507	Jan. 2, 1963
418	62°39'S	56°10'W	311-426	Jan. 2, 1963
426	62°27'S	57°58'W	809-1116	Jan. 5, 1963
428	62°41'S	57°51'W	662-1120	Jan. 5, 1963
437	62°50'S	60°40'W	267-311	Jan. 9, 1963
444	62°56'S	62°02'W	732-750	Jan. 11, 1963
499	62°06'S	45°08'W	485	Feb. 20, 1963
556	51°53'S	56°40'W	849-869	March 14, 1963
558	51°58'S	56°38'W	646-845	March 14, 1963
598	58°13'S	25°50'W	2384-2416	May 3, 1963
671	54°41'S	38°38'W	220-320	Aug. 23, 1963
678	54°49'S	38°01'W	732-814	Aug. 24, 1963
684	54°55'S	38°05'W	595-677	Aug. 25, 1963
740	56°06'S	66°19'W	384-494	Sept. 18, 1963
958	52°56'S	75°00'W	92-101	Feb. 5, 1964
970	54°59'S	64°53'W	586-641	Feb. 11, 1964
973	55°18'S	64°47'W	1922-2229	Feb. 11, 1964
974	53°32'S	64°57'W	119-124	Feb. 12, 1964
976	52°35'S	65°08'W	128	Feb. 13, 1964
977	52°32'S	63°53'W	229	Feb. 13, 1964
980	52°30'S	67°14'W	82	Feb. 14, 1964
992	61°19'S	56°28'W	403	March 13, 1964
993	61°25'S	56°30'W	300	March 13, 1964
997	61°44'S	55°56'W	769	March 14, 1964
1002	62°40'S	54°45'W	265	March 15, 1964
1054	60°37'S	29°59'W	824-1080	April 2, 1964
1067	59°57'S	34°41'W	1098-1153	April 8, 1964
1079	61°26'S	41°55'W	593-598	April 13, 1964
1081	60°35'S	40°44'W	631-641	April 13, 1964
1082	60°50'S	42°55'W	298-302	April 14, 1964
1083	60°51'S	42°57'W	284	April 14, 1964
1084	60°22'S	46°50'W	298-403	April 15, 1964
1088	60°49'S	53°28'W	587-589	April 17, 1964
1089	60°47'S	53°30'W	641	April 17, 1964
1148	65°14'S	117°30'W	4840-4850	June 15, 1964
1284	43°13'W	97°43'W	156-174	Sept. 13, 1964
1344	54°49'S	129°49'W	586-824	Nov. 7, 1964
1345	54°50'S	129°48'W	915-1153	Nov. 7, 1964
1346	54°49'S	129°48'W	549	Nov. 7, 1964
1398	44°00'S	178°06'W	430	Nov. 29, 1964
1403	41°42'S	175°29'E	946-951	Jan. 31, 1965
1411	51°00'S	162°01'E	333-371	Feb. 8, 1965
1412	51°07'S	162°03'E	1647-1665	Feb. 8, 1965
1414	52°17'S	160°40'E	659-798	Feb. 9, 1965
1416	53°45'S	159°09'E	787-842	Feb. 9, 1965
1422	56°19'S	158°29'E	833-842	Feb. 12, 1965
17-6	52°10'S	142°10'W	2304-2328	March 21, 1965
20-134	59°48'S	144°45'E	3200-3259	Oct. 3, 1965
21-282	53°04'S	75°43'W	1896-1920	Jan. 5, 1966
21-283	53°13'S	75°41'W	1500-1666	Jan. 5, 1966
1521	54°09'S	52°08'W	419-483	Jan. 30, 1966
1535	53°51'S	37°38'W	97-101	Feb. 7, 1966
1536	54°29'S	39°22'W	659-686	Feb. 8, 1966
1545	61°04'S	39°55'W	2355-2897	Feb. 11, 1966
1592	54°43'S	55°30'W	1647-2044	March 14, 1966
1596	54°39'S	57°09'W	124	March 14, 1966
1605	52°53'S	74°05'W	522-544	April 1, 1966
1691	53°56'S	140°19'W	362-567	May 14, 1966
1718	38°27'S	168°07'W	531-659	July 12, 1966
25-236	46°04'S	83°55'W	298	Oct. 9, 1966
1814	38°58'S	172°59'E	124	Nov. 30, 1966
1816	40°03'S	168°02'E	90-100	Dec. 2, 1966
1818	40°15'S	168°16'E	913-315	Dec. 2, 1966
1846	43°54'S	167°43'E	1693	Dec. 17, 1966

TABLE A1. (continued)

Station	Latitude	Longitude	Depth, m	Date
		USNS _Eltanin_ (continued)		
1848	41°35'S	175°00'E	256-490	Dec. 19, 1962
1851	49°40'S	178°53'E	476-540	Jan. 3, 1967
1870	71°17'S	171°33'E	659-714	Jan. 14, 1967
1871	71°23'S	171°12'E	351-357	Jan. 14, 1967
1878	72°57'S	171°35'E	573-576	Jan. 15, 1967
1880	73°32'S	171°26'E	518-545	Jan. 16, 1967
1883	73°59'S	170°41'E	598-613	Jan. 16, 1967
1885	74°30'S	170°10'E	311-328	Jan. 16, 1967
1898	76°02'S	178°22'E	485-490	Jan. 20, 1967
1916	77°33'S	174°43'E	728	Jan. 25, 1967
1922	75°32'S	178°50'W	474-496	Jan. 26, 1967
1924	75°10'S	176°13'W	728-732	Jan. 27, 1967
1926	74°53'S	175°10'W	2143-2154	Jan. 27, 1967
1930	74°19'S	176°39'W	831-836	Jan. 28, 1967
1931	73°56'S	178°56'W	399-401	Jan. 29, 1967
1933	73°22'S	177°37'E	465-474	Jan. 30, 1967
1944	67°23'S	180°00'E	516-595	Feb. 2, 1967
1957	64°59'S	160°36'E	2836-2864	Feb. 7, 1967
1981	47°21'S	147°52'E	910-915	Feb. 24, 1967
1983	47°11'S	147°47'E	1028-1034	Feb. 24, 1967
1989	53°29'S	169°48'E	589-594	Jan. 1, 1968
1995	72°03'S	172°38'E	342-360	Jan. 10, 1968
1996	72°05'S	172°08'E	348-352	Jan. 10, 1968
1997	72°00'S	172°28'E	523-528	Jan. 10, 1968
2002	72°18'S	177°35'E	2005-2010	Jan. 11, 1968
2005	73°02'S	176°54'E	864-870	Jan. 12, 1968
2006	73°02'S	176°48'E	861	Jan. 12, 1968
2007	73°05'S	173°59'E	339-343	Jan. 12, 1968
2016	73°59'S	176°11'E	581-586	Jan. 14, 1968
2018	74°01'S	178°53'E	256-258	Jan. 14, 1968
2021	73°49'S	178°13'W	495-503	Jan. 15, 1968
2022	73°51'S	178°15'W	485	Jan. 15, 1968
2025	75°24'S	174°10'W	1225-1240	Jan. 16, 1968
2031	74°39'S	172°18'E	535	Jan. 17, 1968
2045	76°00'S	176°48'E	566-569	Jan. 20, 1968
2063	78°17'S	177°58'W	636	Jan. 26, 1968
2065	78°23'S	173°06'W	473-475	Jan. 26, 1968
2068	78°24'S	169°00'W	562-564	Jan. 27, 1968
2072	76°23'S	163°28'W	509	Jan. 29, 1968
2075	76°25'S	170°24'W	568	Jan. 30, 1968
2079	75°30'S	173°17'W	1320-1335	Jan. 31, 1968
2082	75°50'S	173°08'W	476	Jan. 31, 1968
2085	77°32'S	172°32'W	468-482	Feb. 1, 1968
2088	76°58'S	171°07'W	430-433	Feb. 2, 1968
2092	76°00'S	168°49'W	526	Feb. 3, 1968
2097	76°08'S	165°04'W	494-498	Feb. 4, 1968
2099	77°02'S	166°44'W	408-415	Feb. 4, 1968
2104	77°33'S	163°02'W	606-638	Feb. 5, 1968
2106	77°00'S	161°57'W	525-537	Feb. 6, 1968
2108	74°55'S	174°12'W	2022-2060	Feb. 7, 1968
2115	73°17'S	177°03'W	1287-1308	Feb. 10, 1968
2117	73°02'S	178°06'W	595-600	Feb. 10, 1968
2119	73°05'S	180°00'	567	Feb. 11, 1968
2120	73°04'S	178°03'E	570	Feb. 11, 1968
2124	71°38'S	172°00'E	606-622	Feb. 12, 1968
2125	71°22'S	170°43'E	160-164	Feb. 13, 1968
2143	49°51'S	178°35'E	2010-2100	Feb. 26, 1968
5761	76°01'S	179°50'E	388-399	Feb. 8, 1972
5762	76°02'S	179°57'W	347-358	Feb. 9, 1972
5765	76°07'S	170°12'W	71-87	Feb. 10, 1972
		ARA _Islas Orcadas_		
575-5	50°51'S	43°03'W	1350-1361	May 9, 1975
575-6	51°02'S	42°48'W	1480-1545	May 9, 1975

TABLE A1. (continued)

Station	Latitude	Longitude	Depth, m	Date
		ARA Islas Orcadas (continued)		
575-8	53°36'S	37°35'W	254-366	May 11, 1975
575-10	53°48'S	37°27'W	165-234	May 12, 1975
575-11	53°38'S	38°02'W	132-143	May 12, 1975
575-12	53°38'S	37°55'W	130-137	May 13, 1975
575-13	53°44'S	38°00'W	128-137	May 13, 1975
575-14	53°42'S	37°57'W	144-150	May 14, 1975
575-17	53°36'S	38°03'W	122-124	May 14, 1975
575-30	53°51'S	36°19'W	185-205	May 19, 1975
575-34	54°42'S	34°51'W	563-598	May 19, 1975
575-52	57°39'S	26°27'W	415-612	May 26, 1975
575-53	57°41'S	26°22'W	355-468	May 26, 1975
575-65	56°44'S	26°59'W	302-375	May 31, 1975
575-66	56°43'S	27°00'W	121-228	May 31, 1975
575-67	56°45'S	27°03'W	137-155	May 31, 1975
575-70	56°24'S	27°25'W	161-210	June 2, 1975
575-89	54°44'S	37°11'W	225-265	June 7, 1975
575-90	54°51'S	37°24'W	223-227	June 7, 1975
575-91	55°01'S	37°43'W	494-501	June 7, 1975
575-93	54°39'S	38°51'W	261-270	June 9, 1975
876-107	60°27'S	46°23'W	102-108	Feb. 16, 1976
876-108	60°26'S	46°24'W	152-159	Feb. 16, 1976
876-110	60°28'S	46°27'W	115-132	Feb. 16, 1976
876-113	60°30'S	46°23'W	124-128	Feb. 17, 1976
876-114	60°30'S	46°43'W	128-130	Feb. 17, 1976
876-118	62°02'S	43°06'W	759-857	Feb. 20, 1976
876-124	61°18'S	44°24'W	278-285	Feb. 22, 1976
876-126	61°17'S	44°29'W	283-305	Feb. 22, 1976
876-127	61°16'S	44°26'W	287-289	Feb. 22, 1976
		R/V Hero		
691-20	65°35'S	67°19'W	161	Feb. 8, 1969
702-465	62°57'S	60°50'W	154	March 28, 1970
712-656	54°48'S	64°42'W	18	April 26, 1971
715-694	54°47'S	64°42'W	9	Oct. 17, 1971
715-875	54°55'S	64°00'W	771-903	Oct. 27, 1971
715-885	54°55'S	64°09'W	493-511	Oct. 30, 1971
715-895	55°00'S	64°50'W	438-548	Nov. 3, 1971
715-902	54°29'S	64°40'W	116-120	Nov. 6, 1971
715-903	54°34'S	64°40'W	84-85	Nov. 6, 1971
721-849	64°47'S	64°07'W	120-165	Jan. 26, 1972
721-1081	67°42'S	70°15'W	500-670	Jan. 26, 1972
721-1084	67°05'S	69°22'W	460-500	Jan. 26, 1972
721-1102	64°02'S	64°07'W	60-90	March 1, 1972
721-1110	64°54'S	64°47'W	460-500	March 4, 1972
731-1812	64°52'S	63°40'W	280-300	Feb. 19, 1973
731-1842	65°30'S	67°31'W	180	Feb. 24, 1973
731-1844	67°15'S	70°10'W	450	Feb. 25, 1973
731-1865	65°30'S	64°36'W	60-125	March 2, 1973
731-1940	64°56'S	63°43'W	220-270	March 9, 1973
731-1947	65°00'S	63°28'W	204-250	March 11, 1973
		Yelcho		
2-11	63°40'S	64°10'W	241	March 5, 1962
		USS Atka (Deep Freeze III)		
23	72°06'S	172°15'E	392	Jan. 12, 1958
		USS Burton Island (Deep Freeze III)		
3	72°08'S	172°10'E	433	Jan. 13, 1958
		USS Edisto (Deep Freeze IV)		
7	23°01'S	40°43'W	850	Jan. 2, 1959
15	71°55'S	15°35'W	1280	Jan. 23, 1959
16	71°45'S	15°36'W	1271	Jan. 23, 1959

TABLE A1. (continued)

Station	Latitude	Longitude	Depth, m	Date
	USS Edisto (Deep Freeze IV) (continued)			
20	77°40'S	35°30'W	384	Jan. 28, 1959
21	77°40'S	35°35'W	412	Jan. 30, 1959
28	65°08'S	66°04'W	135	March 22, 1959
31	66°20'S	67°47'W	326	March 25, 1959
36	65°39'S	66°17'W	307	April 5, 1959
49	40°40'S	57°10'W	85	April 5, 1959
50	39°10'S	56°47'W	82	April 15, 1959
	USS Staten Island			
21	76°30'S	156°19'W		Dec. 26, 1960
	USS Glacier (Deep Freeze IV)			
15	73°59'S	168°29'E	366	Dec. 1, 1958
	USCGC Glacier			
1	74°07'S	39°38'W	731	Feb. 6, 1968
11	74°00'S	54°56'W	438	Feb. 12, 1968
	USCGC Eastwind			
2	77°39'S	166°16'E	315	Feb. 19, 1960
3	77°42'S	166°20'E	432	Feb. 19, 1960
	USCGC Eastwind (Cruise 66)			
4	67°53'S	69°11'W	302	Jan. 24, 1966
6	64°51'S	63°15'W	104-146	Jan. 29, 1966
9	62°43'S	62°18'W	549-558	Feb. 1, 1966
16	63°17'S	59°45'W	174	Feb. 5, 1966
23	60°27'S	45°57'W	210	Feb. 9, 1966
28	60°48'S	44°14'W	188-192	Feb. 11, 1966
32	63°58'S	53°10'W	549	Feb. 14, 1966
35	62°12'S	54°25'W	417	Feb. 16, 1966
36	61°16'S	54°45'W	293-295	Feb. 17, 1966
37	61°12'S	54°44'W	110	Feb. 17, 1966
38	61°15'S	54°48'W	174-192	Feb. 17, 1966
39	61°20'S	55°01'W	722-741	Feb. 17, 1966
	USCGC Westwind			
4	77°42'S	41°04'W	796	Nov. 1, 1958
	USFCS Albatross			
2781	51°52'S	73°41'W	637	Feb. 4, 1888
2782	51°12'S	74°14'W	472	Feb. 6, 1888
2785	48°09'S	74°36'W	821	Feb. 8, 1888
3827	*		584-679	April 1, 1902
4397	33°10'N	121°42'W	4016-4074	April 1, 1904
	R/V Vema			
14-2	41°49'S	64°20'W	129	Jan. 31, 1958
14-6	46°48'S	62°47'W	105	Feb. 4, 1958
14-12	52°32'S	61°15'W	361	Feb. 10, 1958
14-14	54°23'S	65°35'W	75	Feb. 19, 1958
14-16	52°22'S	65°45'W	116	Feb. 20, 1958
14-18	52°43'S	62°25'W	307	Feb. 21, 1958
15-99	54°08'S	62°54'W	119	March 3, 1959
15-102	52°53'S	65°35'W	108	March 5, 1959
15-103	53°12'S	65°30'W	106	March 5, 1959
15-108	54°10'S	64°19'W	110	March 6, 1959
15-109	54°11'S	62°36'W	403	March 7, 1959
15-110	54°10'S	63°20'W	284	March 7, 1959
15PD-3	54°12'S	62°36'W	399	March 7, 1959
15PD-4	54°10'S	63°20'W	284	March 7, 1959
15PD-9	56°28'S	66°54'W	406	March 13, 1959
15PD-10	56°39'S	66°53'W	1137	March 13, 1959
15-132	39°58'S	54°50'W	1912	April 3, 1959

TABLE A1. (continued)

Station	Latitude	Longitude	Depth, m	Date
		R/V Vema (continued)		
16-39	50°53'S	62°35'W	157	May 19, 1960
17-11	43°25'S	75°05'W	152	March 23, 1961
17-14	47°01'S	75°44'W	1201	March 24, 1961
17-39	53°47'S	70°18'W	267	April 3, 1961
17-57	54°57'S	63°04'W	1904	May 8, 1961
17-59	54°54'S	60°27'W	432	May 10, 1961
17-61	54°44'S	55°39'W	1814-1919	May 11, 1961
17-74	41°27'S	59°33'W	71	May 23, 1961
17-76	41°57'S	59°03'W	81	May 23, 1961
17-88	45°11'S	60°55'W	110	June 11, 1961
17-89	45°02'S	61°18'W	102	June 11, 1961
17-90	44°53'S	61°43'W	99	June 11, 1961
17-97	44°29'S	60°59'W	101	June 13, 1961
17-100	44°23'S	59°53'W	166-177	June 13, 1961
17-101	38°13'S	55°19'W	450-454	June 19, 1961
18-8	36°06'S	53°18'W	278-282	Feb. 4, 1962
18-12	47°09'S	60°38'W	424-428	
18-13	47°10'S	61°02'W	135	Feb. 17, 1962
18-14	47°13'S	61°30'W	130-132	Feb. 17, 1962
18-16	47°30'S	62°39'W	123	Feb. 18, 1962
18-32	63°47'S	66°07'W	672-726	March 3, 1962
		NZOI		
A-537	77°30'S	165°12'E	534-574	Feb. 17, 1960
A-625	75°00'S	163°59'E	460-520	Feb. 5, 1961
A-706	47°42'S	178°43'E	311	Nov. 4, 1962
A-740	49°41'S	178°40'E	315	Nov. 9, 1962
A-898	43°22'S	177°17'E	230	
B-314	39°22'S	171°50'E	230-251	Oct. 25, 1960
B-319	39°04'S	172°22'E	638	Oct. 26, 1960
C-410	41°34'S	174°37'E	475	May 4, 1960
C-527	32°30'S	179°12'W	508	Sept. 18, 1960
C-618	43°52'S	175°20'E	625-690	May 2, 1961
C-633	39°16'S	171°54'E	340	May 27, 1961
C-642	39°16'S	171°53'E	350-400	May 28, 1961
C-734	53°55'S	158°55'E	366	Nov. 25, 1961
D-6	55°29'S	158°32'E	415	April 20, 1963
D-145	48°43'S	167°27'E	357	Jan.14, 1964
D-149	49°11'S	166°51'E	448	
D-159	49°01'S	164°30'E	713	Jan.17, 1964
D-160	48°56'S	166°04'E	713	Jan. 18, 1964
D-166	49°50'S	163°59'E	658	Jan. 19, 1964
D-175	50°38'S	167°38'E	421	Jan. 21, 1964
D-176	51°06'S	167°49'E	216-582	Jan. 21, 1964
D-177	51°25'S	167°50'E	640	Jan. 21, 1964
D-179	51°25'S	167°21'E	611	Jan. 22, 1964
D-207	50°04'S	171°23'E	510	Jan. 25, 1964
		Walther Herwig		
215/66	32°00'S	50°00'W	500	June 9, 1966
244/66	36°51'S	54°01'W	800	June 14, 1966
267/66	40°00'S	56°02'W	520	June 19, 1966
268/66	39°56'S	55°58'W	600	June 19, 1966
285/66	42°19'S	58°01'W	825	June 21, 1966
311/66	47°01'S	60°43'W	310	June 25, 1966
324/66	50°01'S	57°16'W	410	June 28, 1966
325/66	50°00'S	56°49'W	515	June 29, 1966
330/66	51°00'S	56°24'W	530	June 29, 1966
336/66	51°57'S	56°42'W	600	June 30, 1966
357/66	52°49'S	63°13'W	335	July 11, 1966
359/66	52°49'S	62°52'W	330	July 12, 1966
360/66	52°24'S	62°21'W	295	July 12, 1966
361/66	51°55'S	61°50'W	200	July 12, 1966
64/68	30°03'S	47°44'W	800	Feb. 27, 1968

TABLE A1. (continued)

Station	Latitude	Longitude	Depth, m	Date
Walther Herwig (continued)				
142/71	42°06'S	57°55'W	708-765	Jan. 4, 1971
191/71	46°13'S	59°49'W	805	Jan. 17, 1971
197/71	48°13'S	60°10'W	500	Jan. 19, 1971
269/71	53°02'S	60°00'W	442	Feb. 9, 1971
270/71	53°00'S	60°00'W	375	Feb. 9, 1971
328/71	42°52'S	58°38'W	1200	Feb. 22, 1971
329/71	41°13'S	56°51'W	1250	Feb. 22, 1971
331/71	41°05'S	57°15'W	775	Feb. 23, 1971
R/V Anton Bruun				
11-88	07°58'S	80°37'W	1005-1124	Oct. 14, 1965
18-714	25°00'S	70°40'W	950	Aug. 16, 1966
Calypso				
171	37°36'S	54°46'W	740	Dec. 29, 1961
172	37°35'S	54°54'W	220-270	Dec. 29, 1961
Golden Hind				
35	35°08'S	139°32'E	732	
BR				
25149	37°35'S	54°55'W	115	

*Northwest of Lae-o ka Laau Lt., Molokai Island, Hawaiian Islands.

Acknowledgments. This work was supported by the Cooperative Systematics and Analyses of Polar Biological Materials program (National Science Foundation grant DPP 76-23979). Thanks are extended to G. Hendler and B. Landrum, both of the Smithsonian Oceanographic Sorting Center, for making the USARP specimens available to me and for supplying additional data on the collection. I am also thankful to the administration of the Smithsonian Institution for allowing me to work at the United States National Museum during this study. I would like to thank the following people who have generously extended to me the use of their collections and facilities or loaned me specimens used in this study: P. F. S. Cornelius (BM), E. Kirsteuer (AMNH), H. Zibrowius (Station Marine d'Endoume, Marseille), K. P. Sebens (MCZ), W. Zeidler (South Australian Museum), J. K. Lowry (Australian Museum), J. P. Chevalier (MNHNP), D. H. H. Kühlmann (Museum für Naturkunde an der Humboldt-Universität, Berlin), and J. Van Goethem (Institut Royal des Sciences Naturelles de Belgique, Brussels). I am very grateful to my wife, Peggy, for her editing, proofreading, and typing of the manuscript and to my father, E. J. Cairns, for his careful editing. I would like to thank the following people for having read the manuscript and offered valuable suggestions: H.

Zibrowius, Station Marine d'Endoume; J. W. Wells, Cornell University; and F. M. Bayer, Smithsonian Institution. The author is a research associate at the Department of Invertebrate Zoology of the Smithsonian Institution, Washington, D. C.

References

Agassiz, A.
   1888 Three cruises of the...Steamer Blake. Bull. Mus. Comp. Zool. Harv., 15: 148-156, text figs. 462-481.
Alcock, A.
   1898 An account of the deep-sea Madreporaria collected by the royal Indian marine survey ship Investigator. 29 pp., 3 pls. Trustees of Indian Museum, Calcutta.
   1902 Report on the deep-sea Madreporaria of the Siboga-Expedition. Siboga Exped. Monogr., 16a: 52 pp., 5 pls.
Avent, R. M., M. E. King, and R. H. Gore
   1977 Topographic and faunal studies of the shelf-edge prominences off the central eastern Florida coast. Int. Revue Ges. Hydrobiol. Hydrogr., 62(2): 185-208, 11 figs.

Beurois, J.
  1975 Etude écologique et halieutique des fonds de pêche et des espèces d'intérêt commercial (langoustes et poissons) des Îles Saint-Paul et Amsterdam (Ocean Indien). Rep. 37: 91 pp. Com. Nat. Fr. des Rech. Antarct., Paris.

Briggs, J. C.
  1974 Marine zoogeography. 475 pp. McGraw-Hill, New York.

Broch, H.
  1927 The Folden fjord. Coelenterata. 2. Tromsø Mus. Arsh., 1(12): 7, 8.

Broderip, W. J.
  1828 Description of Caryophyllia smithii n. sp. J. Zool., 3: 485, 486, pl. 13, figs. 1-6.

Brodie, J. W., and E. W. Dawson
  1965 Morphology of North Macquarie Ridge. Nature, 207(4999): 844, 845, 1 fig.

Bullivant, J. S.
  1967 Ecology of the Ross Sea benthos. N.Z. Dep. Sci. Industr. Res. Bull., 176: 49-75, pls. 11-23.

Cailleux, A.
  1961 Endemicité actuelle et passée de l'Antarctique. C. R. Somm. Seanc. Soc. Biogeogr., no. 332-334: 65-71.

Cairns, S. D.
  1979 The deep-water Scleractinia of the Caribbean Sea and adjacent waters. Stud. Fauna Curacao, 57(180): 341 pp., incl. 40 pls.

David, T. W. E., and R. E. Priestley
  1914 Glaciology, physiography, stratigraphy, and tectonic geology of South Victoria. Land. Rep. Scient. Invest. Br. Antarct. Exped. 1907-1909, Geol., 1: 319 pp.

Dawson, E. W.
  1970 Faunal relationships between the New Zealand Plateau and the New Zealand sector of Antarctica based on echinoderm distribution. N.Z. J. Mar. Freshwat. Res., 4: 126-140.

Dell, R. K.
  1972 Antarctic benthos. Adv. Mar. Biol., 10: 1-216.

Dennant, J.
  1906 Madreporaria from the Australian and New Zealand coasts. Trans. R. Soc. S. Aust., 30: 151-165, pls. 5, 6.

Duchassaing, P., and J. Michelotti
  1864 Supplément au mémoire sur les coralliaires des Antilles. Mem. Acad. Sci. Torino, Ser. 2, 23: 97-206 (reprint numbered 1-112), 11 pls.

Duncan, P. M.
  1873 A description of the Madreporaria dredged up during the expedition of the H.M.S. Porcupine in 1869 and 1870. Part 1. Trans. Zool. Soc. Lond., 8(5): 303-344, pls. 39-49.
  1876 Notice of some deep-sea and littoral corals from the Atlantic Ocean, Caribbean, Indian, New Zealand, Persian Gulf, and Japan &c Seas. Proc. Zool. Soc. Lond., pp. 428-442, pls. 38-41.

Eguchi, M.
  1965 On some deep water corals from the Antarctic Sea. JARE Scient. Rep., Ser. E, no. 28: 1-12, pls. 1, 2.
  1968 The hydrocorals and scleractinian corals of Sagami Bay collected by His Majesty the Emperor of Japan. xv + 221 pp., 70 pls. Mazuren, Tokyo.

Faustino, L. A.
  1927 Recent Madreporaria of the Philippine Islands. Philipp. Bur. Sci. Monogr., 22: 310 pp., 100 pls.

Fowler, G. H.
  1885 The anatomy of the Madreporaria. Part 1. Q. Jl Microsc. Sci., 25: 577-597, pls. 40-42.

Gardiner, J. S.
  1904 The turbinolid corals of South Africa, with notes on their anatomy and variations. Mar. Invest. S. Afr., 3(4): 93-129, 3 pls.
  1913 The corals of the Scottish National Antarctic Expedition. Trans. R. Soc. Edinb., 49(3): 687-689.
  1929a Turbinolidae and Eupsammidae. Br. Antarct. Terra Nova Exped. 1910 (Zool.), 5(4): 121-130, 1 pl.
  1929b Corals of the genus Flabellum from the Indian Ocean. Rec. Indian Mus., 31(4): 301-310, pl. 13.
  1939 Madreporarian corals, with an account of variation in Caryophyllia. Discovery Rep., 18: 323-338, pls. 20, 21.

Gardiner, J. S., and P. Waugh
  1938 The flabellid and turbinolid corals. Scient. Rep. John Murray Exped., 5(7): 167-202, pls. 1-7, 6 text figs.
  1939 Madreporaria excluding the Flabellidae and Turbinolidae. Scient. Rep. John Murray Exped., 6(5): 225-242, pls. 1, 2, 3 text figs.

Gravier, C.
  1914a Sur une espèce nouvelle de madréporaire (Desmophyllum antarcticum). Deuxième Expédition Antarctique Française. Bull. Mus. Natn. Hist. Nat. Paris, 20: 236-238.
  1914b Madréporaires. Deuxième Expédition Antarctique Française. Pp. 119-133, 1 pl. Masson, Paris.
  1920 Madréporaires provenant des campagnes des yachts Princesse Alice et Hirondelle II (1893-1913). Result. Camp. Scient. Prince Albert I, 55: 123 pp., 16 pls.

Hedgpeth, J. W.
  1969 Distribution of selected groups of marine invertebrates in waters south of 35°S latitude. Introduction to Antarctic zoogeography. Antarctic map folio ser., 11: 1-9, 14 figs. American Geographical Society, New York.

Hoffmeister, J. E.
  1933 Report on the deep-sea corals obtained by the F.I.S. Endeavour on the coasts of New South Wales, Victoria, South Australia, and Tasmania. Zool. (Biol.) Results Fish. Exp. Endeavour, 6(1): 1-16, 4 pls.

Jourdan, E.
  1895 Zoanthaires provenant des campagnes du yacht l'Hirondelle. Result. Camp. Scient. Prince Albert I, 8: 36 pp., 2 pls.

Jungersen, H.
  1916 Alcyonarian and madreporarian corals in the Museum of Bergen, collected by the

Fram expedition 1898-1902, and by the
Michael Sars 1900-1906. Bergens Mus.
Aarb., 1915-1916, no. 6: 1-44.

Keller, N. B.
1974   New data about some species of madrepor-
       arian corals of the genus Flabellum.
       Trudy Inst. Okeanol., 98: 199-212, pls.
       1-7.

1975   Ahermatypic madreporarian corals of the
       Caribbean Sea and the Gulf of Mexico.
       Trudy Inst. Okeanol., 100: 174-187, 2 pls.

1976   The deep-sea madreporarian corals of the
       genus Fungiacyathus from the Kurile-
       Kamchatka, Aleutian trenches and other
       regions of world ocean. Trudy Inst.
       Okeanol., 99: 31-44, 3 pls.

1977   New species of the genus Leptopenus and
       some peculiarities of the deep-sea
       ahermatypic corals. Trudy Inst. Okeanol.,
       108: 37-43, 3 figs., 1 pl.

Knox, G. A.
1970   Biological oceanography of the South
       Pacific. In W. S. Wooster (Ed.), Scien-
       tific exploration of the South Pacific.
       pp. 155-182. National Academy of Sci-
       ences, Washington, D. C.

Laborel, J.
1970   Les peuplements de madréporaires des
       côtes tropicales du Brésil. Ann. Univ.
       Abidjan, Ser. E., 2(3): 261 pp., 71 figs.

1971   Madréporaires et hydrocoralliaires réci-
       faux des côtes brésiliennes. Result.
       Scient. Camp. Calypso, no. 9, 36(25):
       171-229, 8 pls., 6 text figs. (Also in
       Ann. Inst. Oceanogr. Monaco, 47: 171-229,
       8 pls.)

Linnaeus, C.
1758   Systema naturae...I. 10th ed., 824 pp.,
       Stockholm.

Marion, A. F.
1906   Etude des coelentérés atlantiques recueil-
       lis par la commission de dragages de
       l'aviso 'Le Travailleur' durant les
       campagnes 1880 et 1881. In Expédition
       Scientifique Travailleur et Talisman. Pp.
       103-151, pls. 11-17. Masson, Paris.

Menzies, R. J., R. Y. George, and G. T. Rowe
1973   Abyssal environment and ecology of the
       world oceans. xxiii + 488 pp. Wiley, New
       York.

Milne Edwards, H., and J. Haime
1848   Recherches sur les polypiers. Mém. 2.
       Monographie des turbinolides. Ann. Sci.
       Nat., Ser. 3, 9: 211-344, pls. 7-10.

1850   A monograph of the British fossil corals.
       Part 1. Introduction: Corals of the Ter-
       tiary and Cretaceous formations. lxxxv +
       171 pp., 11 pls. Palaeontographical Soci-
       ety, London.

1857   Histoire naturelle des coralliaires ou
       polypes proprement dits. Vol. 2, 633 pp.
       Roret, Paris.

Moseley, H. N.
1876   Preliminary report to Professor Wyville
       Thomson...on the true corals dredged by
       the H.M.S. Challenger in deep water
       between the dates Dec. 30th, 1870, and
       August 31st, 1875. Proc. R. Soc., 24:
       544-569, 1 fig.

1881   Report on certain hydroid, alcyonarian
       and madreporarian corals procured during
       the voyage of the H.M.S. Challenger, in

the years 1873-1876. Part 3. On the deep-
       sea Madreporaria. Rep. Scient. Results
       Challenger (Zool.), 2: 127-208, 16 pls.,
       21 text figs.

Niino, H.
1958   On the bottom deposits of the sea around
       Cape Cook, Prince Harald Coast, Ant-
       arctica. J. Tokyo Univ. Fish., spec. ed.,
       1(2): 250-257, 2 pls.

Nordgaard, O.
1929   Faunistic notes on marine evertebrates.
       6. On the distribution of some madrepor-
       arian corals in northern Norway. K. Norske
       Vidensk. Selsk. Forh., 2: 102-105.
       105.

Pallas, P. S.
1766   Elenchus zoophytorum. xvi + 28 + 451 pp.
       Hague-Comitum, Netherlands.

Pawson, D. L.
1969   Distribution of selected groups of marine
       invertebrates in waters south of 35°S
       latitude: Echinoidea. Antarctic map folio
       ser., 11: 38-41, pl. 23. American Geo-
       graphic Society, New York.

Pax, F.
1910   Die Steinkorallen der deutschen Südpolar-
       Expedition 1901-1903. Dt. Sudpol.-Exped.,
       12(1): 63-76, 2 pls.

Podoff, N.
1976   Microstructure of modern deep-water corals
       (Flabellidae and Parasmilinae). M.A.
       thesis, 104 pp., 14 pls. State Univ. of
       New York, Binghamton.

Pourtalès, L. F.
1868   Contributions to the fauna of the Gulf
       Stream at great depths. Bull. Mus. Comp.
       Zool. Harv., Ser. 2, 1(7): 121-141.

1878   Reports on the results of dredging...by
       the Blake. Bull. Mus. Comp. Zool. Harv.,
       5(9): 197-212, 1 pl.

1880   Reports on the results of dredging...by
       the Blake. Report on the corals and
       Antipatharia. Bull. Mus. Comp. Zool.
       Harv., 6(4): 95-120, 3 pls.

Ralph, P. M.
1948   Some New Zealand corals, N.Z. Sci. Rev.,
       6(6): 107-110, 4 figs.

Ralph, P. M., and D. F. Squires
1962   The extant scleractinian corals of New
       Zealand. Zoology Publs Vict. Univ.
       Wellington, 29: 1-19, 8 pls., 1 text fig.

Rehberg, H.
1892   Neue and wenig bekannte Korallen. Abh.
       Naturw. Ver. Hamburg, 12: 1-50, 4 pls.

Ridgway, N. M.
1968   New Zealand region: Sea water temperatures
       at the ocean floor. N.Z. Oceanogr. Inst.
       Chart. Misc. Ser., map 16.

Sars, G. O.
1872   On some remarkable forms of animal life
       from great depths off the Norwegian coast.
       2. University Program for the first half-
       year 1869. 82 pp., 6 pls. Brøgger &
       Christie, Oslo.

Sars, M.
1869   Remarks on the distribution of animal
       life in the depths of the sea. Ann. Mag.
       Nat. Hist., Ser. 4, 3: 423-441.

Sorauf, J. E., and N. Podoff
1977   Skeletal structure in deep water aherma-
       typic corals. Mem. Bur. Rech. Geol. Min-
       iers, 89: 2-11, 4 pls. (Deuxième sympo-

sium international sur les coraux et re-
cifs coralliers fossiles, Paris, 1975.)

Speden, I. G.
1962 Fossiliferous Quaternary marine deposits
in the McMurdo Sound region, Antarctica.
N.Z. Jl Geol. Geophys., 5: 746-777, 16
figs.

Squires, D. F.
1958 The Cretaceous and Tertiary corals of New
Zealand. N.Z. Geol. Surv. Paleont. Bull.,
29: 1-107, 16 pls.
1959 Deep sea corals collected by the Lamont
Geological Observatory. 1. Atlantic
corals,. Am. Mus. Novit., no. 1965: 42
pp., 24 figs.
1960 Scleractinian corals from the Norfolk
Island cable. Rec. Auckland Inst. Mus.,
5(3, 4): 195-201, pls. 33-35.
1961 Deep sea corals collected by the Lamont
Geological Observatory. 2. Scotia Sea
corals. Am. Mus. Novit., no. 2046: 48
pp., 31 figs.
1962a Deep sea corals collected by the Lamont
Geological Observatory. 3. Larvae of the
Argentine scleractinian coral Flabellum
curvatum Moseley. Am. Mus. Novit., no.
2078: 11 pp., 3 figs.
1962b The fauna of the Ross Sea. 2. Scleractinian corals. N.Z. Dep. Sci. Industr.
Res. Bull., 147: 1-28, pls. 1, 2, text
figs. 1-8.
1963a Madreporas rhizángidas, fósiles y
vivientes de la Argentina. Neotropica,
9(28): 9-16, figs. 1-11.
1963b Flabellum rubrum (Quoy and Gaimard). N.Z.
Dep. Sci. Industr. Res. Bull., 154: 43
pp., 2 pls.
1964a Biological results of the Chatham Islands
1954 Expedition. Part 6. Scleractinia.
N.Z. Dep. Sci. Industr. Res. Bull.,
139(6): 31 pp., 4 pls.
1964b New stony corals (Scleractinia) from
northeastern New Zealand. Rec. Auckland
Inst. Mus., 6(1): 9 pp., 2 pls.
1964c The Southern Ocean: A potential for coral
studies. Report for 1963, pp. 447-459,
pls. 1-4, text figs. 1-4. Smithson. Inst.,
Washington, D. C.
1965a A new record for Leptopenus, a rare deepwater coral. Nature, 207(4999): 878-879,
1 fig.
1965b Deep-water coral structure on the Campbell
Plateau, New Zealand. Deep Sea Res., 12:
785-788, 2 figs.
1967 The evolution of the deep-sea family
Micrabaciidae. Stud. Trop. Oceanogr., 5:
502-510.
1969 Distribution of selected groups of marine
invertebrates in waters south of 35°S
latitude: Scleractinia. Antarctic map
folio ser., 11: 15-18, pl. 6. American
Geographical Society, New York.

Squires, D. F., and I. W. Keyes
1967 The marine fauna of New Zealand: Scleractinian corals. N.Z. Dep. Sci. Industr.
Res. Bull., 185: 1-46, 4 pls., 7 text
figs.

Squires, D. F., and P. M. Ralph
1965 A new scleractinian coral of the genus
Flabellum from New Zealand, with a new
record of Stephanocyathus. Proc. Biol.
Soc. Wash., 78: 259-264, 1 pl.

Studer, T.
1878 Übersicht der Steinkorallen aus der
Familie der Madreporaria aporosa,
Eupsammina and Turbinarina, welche auf
der Reise S.M.S. Gazelle um die Erde
gesammelt wurden. Monatber. Kon. Preuss.
Akad. Wiss. Berlin, Nov. 1, 1877: pp.
625-655, 4 pls.

Thomson, J. A.
1931 Alcyonarians and solitary corals. Rep.
Scient. Results Michael Sars N. Atlant.
Deep Sea Exped., 5: 1-10, 2 pls.

Thomson, J. A., and N. Rennet
1931 Alcyonaria, Madreporaria and Antipatharia.
Scient. Rep. Australas. Antarct. Exped.,
Ser. C, 9(3): 1-46, pls. 8-14.

Vaughan, T. W.
1906a A new species of Coenocyathus from
California and the Brazilian astrangid
corals. Proc. U.S. Natn. Mus., 30(1477):
847-850, pls. 77, 78.
1906b Reports on the scientific results of the
expedition to the eastern tropical
Pacific...by the U.S. Fish Commission
Steamer Albatross from October, 1904, to
March, 1905. 6. Madreporaria. Bull. Mus.
Comp. Zool. Harv., 50(3): 61-72, pls.
1-10.
1907 Recent Madreporaria of the Hawaiian
Islands and Laysan. Bull. U.S. Natn. Mus.,
59: 427 pp., 96 pls.

Verrill, A. E.
1869 Notes on Radiata in the museum of Yale
College. No. 6. Review of the corals and
polyps of the west coast of America.
Trans. Conn. Acad. Arts Sci., 1: 377-558,
pls. 4-10.
1882 Notice of the remarkable marine fauna
occupying the outer banks off the southern
coast of New England. No. 5. Amer. J.
Arts Sci., Ser. 3, 23: 309-316.
1883 Report on the Anthozoa and on some additional species dredged by the Blake in
1877-79, and by the U.S. Fish Commission
Steamer Fish Hawk in 1880-82. Bull. Mus.
Comp. Zool. Harv., 11: 1-72, pls. 1-8.
1885 Notice of the remarkable marine fauna
occupying the outer banks off the southern
coast of New England. No. 11. Amer. J.
Arts Sci., Ser. 3, 29: 149-157.
1901 Variations and nomenclature of Bermudian,
West Indian, and Brazilian coral reefs,
with notes on various Indo-Pacific corals.
Trans. Conn. Acad. Arts Sci., 11: 63-168,
pls. 10-35.

von Marenzeller, E.
1903 Madreporaria and Hydrocorallia. Result.
Voyage S.Y. Belgica, 7: 1-7, 1 pl.
1904a Steinkorallen. Wiss. Ergebn. Dt. Tiefsee-
Exped. Valdivia, 7: 261-318, pls. 14-18.
1904b Stein- und Hydrokorallen. Bull. Mus.
Comp. Zool. Harv., 43(2): 75-87, 3 pls.

Weisbord, N. E.
1968 Some late Cenozoic stony corals from
northern Venezuela. Bull. Am. Paleont.,
55(246): 281 pp., 12 pls.
1974 Late Cenozoic corals of South Florida.
Bull. Am. Paleont., 66(285): 544 pp., 57
pls.

Wells, J. W.
1936 The nomenclature and type species of some
genera of Recent and fossil corals. Amer.

J. Arts Sci., Ser. 5, 31: 97–134.

1956 Scleractinia. In R. C. Moore (Ed.), Treatise on invertebrate paleontology. Part F. Coelenterata. Pp. 328–444, figs. 223–339. Lawrence, Kansas.

1958 Scleractinian corals. Rep. BANZ Antarct. Res. Exped., Ser. B, 6(11): 257–275, 2 pls.

1966 Evolutionary development in the scleractinian family Fungiidae. Symp. Zool. Soc. Lond., 16: 223–246.

Yabe, H., and M. Eguchi

1932 A study of the Recent deep-water coral fauna of Japan. Proc. Imp. Acad. Japan, 8(8): 387–390.

1936 Deep-water corals from off Owasi, Mie Prefecture. Proc. Imp. Acad. Japan, 12(10): 167–168.

1941a Fossil and Recent Flabellum from Japan. Sci. Rep. Tohoku Imp. Univ., Ser. 2, 22(2): 87–103, pls. 5–8.

1941b Fossil and Recent simple corals from Japan. Sci. Rep. Tohoku Imp. Univ., Ser. 2, 22(3): 105–178, pls. 9–12.

1943 Note on the two hexacorallia Goniocorella dumosa (Alcock) and Bantamia gerthi, gen. et sp. nov. Proc. Imp. Acad. Japan, 19: 494–500, figs. 1–5.

Zibrowius, H.

1973 Révision des espèces actuelles du genre Enallopsammia Michelotti, 1871 et description de E. marenzelleri, nouvelle espèce bathyale à large distribution: Océan Indien et Atlantique Central (madréporaires, Dendrophylliidae). Beaufortia, 21(276): 37–54, 3 pls.

1974a Scléractiniaires des Îles Saint Paul et Amsterdam (sud de l'Océan Indien). Tethys, 5(4): 747–778, 3 pls.

1974b Révision du genre Javania et considérations générales sur les Flabellidae (scléractiniaires). Bull. Inst. Oceanogr. Monaco, 71(1429): 1–48, 5 pls.

1974c Oculina patagonica, scléractiniaire hermatypique introduit en Méditerranée. Helgolander Wiss. Meeresunters., 26(2): 153–173.

1980 Les scléractiniaires de la Méditerranée et de l'Atlantique nordoriental. Mem. Inst. Oceanogr. Monaco, 11: 248 pp., 107 pls.